Green Goals and Greenbacks

Westview Replica Editions

This book is a Westview Replica Edition. The concept of Replica Editions is a response to the crisis in academic and informational publishing. Library budgets for books have been severely curtailed; economic pressures on the university presses and the few private publishing companies primarily interested in scholarly manuscripts have severely limited the capacity of the industry to properly serve the academic and research communities. Many manuscripts dealing with important subjects, often representing the highest level of scholarship, are today not economically viable publishing projects. Or, if they are accepted for publication, they are often subject to lead times ranging from one to three years. Scholars are understandably frustrated when they realize that their first-class research cannot be published within a reasonable time frame, if at all.

Westview Replica Editions are our practical solution to the problem. The concept is simple. We accept a manuscript in camera-ready form and move it immediately into the production process. The responsibility for textual and copy editing lies with the author or sponsoring organization. If necessary we will advise the author on proper preparation of footnotes and bibliography. We prefer that the manuscript be typed according to our specifications, though it may be acceptable as typed for a dissertation or prepared in some other clearly organized and readable way. The end result is a book produced by lithography and bound in hard covers. Initial edition sizes range from 500 to 800 copies, and a number of recent Replicas are already in second printings. We include among Westview Replica Editions only works of outstanding scholarly quality or of great informational value, and we will continue to exercise our usual editorial standards and quality control.

Green Goals and Greenbacks:
State-Level Environmental Review Programs and Their Associated Costs

Stuart L. Hart and Gordon A. Enk

Focusing on the eighteen states that have established Environmental Impact Statement (EIS) requirements similar to those of the National Environmental Policy Act of 1969, Stuart Hart and Gordon
Enk analyze implementation of EIS programs at the
state level, examine costs, and suggest ways to improve efficiency and effectiveness. In their initial
discussion of data collection and methodology, the
authors introduce an EIS Cost Accounting System as
an aid to understanding the form and interrelation-
ship of the costs of the EIS process; they then identify and explain fundamental areas of the EIS process (e.g., policy statements, enforcement mechanisms,
public participation). Though they conclude that,
contrary to often cited criticism, the costs of environmental review are relatively insignificant, they
offer a number of techniques to make the process still
less expensive in terms of both time and dollars.

Stuart L. Hart is presently research associate
in economic and environmental studies at the Institute on Man and Science, where he also served as
environmental planning consultant.

Gordon A. Enk, director of economic and environmental studies at the Institute on Man and Science,
has previously been involved with the U. S. Forest
Service as research associate and with the Department
of Natural Resources in the state of Washington as
research economist.

Green Goals and Greenbacks:

State-Level Environmental Review Programs and Their Associated Costs

Stuart L. Hart and Gordon A. Enk

Routledge
Taylor & Francis Group

LONDON AND NEW YORK

First published 1980 by Westview Press, Inc.

Published 2018 by Routledge
52 Vanderbilt Avenue, New York, NY 10017
2 Park Square, Milton Park, Abingdon, Oxon OX14 4RN

Routledge is an imprint of the Taylor & Francis Group, an informa business

Library of Congress Catalog Card Number: 79-5229

ISBN 13: 978-0-367-02196-2 (hbk)
ISBN 13: 978-0-367-17183-4 (pbk)

Contents

Tables

Figures

Preface

Since the signing of the National Environmental Policy Act (NEPA) on January 1, 1970, eighteen states have initiated their own programs of comprehensive environmental review. These State Environmental Policy Acts (SEPA's), or "NEPA equivalents," are generally patterned after the policies, goals, and processes established by NEPA but, at the same time, demonstrate a great deal of variability in terms of extent as well as style of implementation. They are "equivalent," however, in the sense that they all utilize a similar tool--the environmental impact statement (EIS).

This study focuses on the eighteen states which have established their own EIS requirements through legislation, executive order, or administrative action. Our purpose was to analyze how these states implement their EIS programs, to examine the costs associated with the EIS process, and to suggest ways of making the overall process more efficient and effective. The report is intended to be of use not only to legislators and decision makers within the eighteen states, but also to those states currently contemplating EIS requirements. It should also serve as a rich information base to a variety of environmental researchers and educators. To these ends, the report has been divided into two parts. Part 1 contains the comparative analysis of state programs as well as all conclusions and recommendations. Part 2 contains detailed supporting material pertaining to the evolution of each SEPA and its associated costs.

Chapter 1 presents a discussion of data collection and methodology and describes the nature of the costs associated with the EIS process. An "EIS Cost Accounting System" is introduced to aid in understanding the concepts of where these costs are in-

curred, what form they take, and how they are interrelated.

Chapter 2 deals with the "fundamental" areas of
the EIS process such as policy statements, elements
of EIS's, enforcement mechanisms, and public participation. Indeed, in contemplating an EIS requirement,
a state is faced with several options concerning the
scope of the program, who will be responsible for it,
and how it will be carried out. The experience of
the eighteen SEPA states in these matters is comparatively analyzed and recommendations are offered
based on our findings.

An often cited criticism of environmental programs in general and EIS requirements in particular
is that such requirements encourage excessive government spending, impede economic growth, and contribute to unemployment. Chapter 3 addresses the complex issue of the costs of the environmental impact
statement process, dealing largely with the public
sector costs of environmental review at the state
level. Problems in assessing EIS-related costs such
as inadequate agency records and cost mixing are discussed followed by a general analysis of state EIS
program coordinating costs, document preparation and
review costs, total costs, and trends in cost and
caseload. Although it is concluded that the costs
of environmental review are relatively insignificant,
a number of techniques are identified which have the
effect of mitigating unnecessary delay, duplication,
and overlap in the EIS process. Following comparative analysis of the innovative techniques being
utilized by the eighteen SEPA states in this regard,
recommendations intended to foster a more efficient
and effective EIS process are offered.

Chapters 4 and 5 contain detailed information
about each state's EIS program structure, evolution,
and cost. Questionnaire data is verified, supplemented, or balanced where appropriate by official
documents such as SEPA legislation, guidelines, regulations, budgets, reports, studies, or personal correspondence. In-depth, personal staff investigation
in the six "focus" states of Chapter 4 resulted in a
level of information warranting separation from that
of the other twelve "brief" states contained in Chapter 5. Finally, Appendix A--the SEPA "Fact Sheets"--
provides a succinct "snapshot" of the current status
of all eighteen SEPA's analyzed in the study.

This report constitutes the second of a three-
phased study of The Economics of Environmental Impact
Statements. The first phase, Beyond NEPA Revisited:
Directions in Environmental Impact Review, published

in March of 1978, focused on the impact of NEPA at
the state level and provides information on the evolution and current status of federal EIS review pro-
cedures--through the A-95 Clearinghouses--in all
fifty states. The final phase, forthcoming in the
summer of 1979, will clarify and develop an approach
to understanding the benefits derived from the EIS
process.

Acknowledgments

This study would not have been possible without the responsive assistance of numerous people in
state government. Appendix D identifies those state
government representatives who gave generously of
their time in assisting our understanding of their
environmental impact statement processes. In particular, we would like to thank Norman Hill of the
California Resources Agency; Scott Warner of the
California Office of Planning and Research; William
Hicks of the Massachusetts Office of Environmental
Affairs; Richard Bates of the Massachusetts Depart-
ment of Environmental Quality Engineering; John
Robertson, Charles Kenow, and Nancy Onkka of the
Minnesota State Planning Agency; Dale McMichael of
the Minnesota Pollution Control Agency; Terence
Curran, Allen Davis, Michael Morandi, and Pat Grady
of the New York State Department of Environmental
Conservation; Peter Buttner of the New York State
Office of Parks and Recreation; Keith Whitenight and
Thayer Broili of the North Carolina Department of
Natural and Economic Resources; Thomas Elwell,
Dennis Lundblad, and Stan Springer of the Washington
State Department of Ecology; and Bernie Chaplin of
the Washington State Department of Transportation.

We would also like to express our gratitude to
Gary Toenniessen and Ralph Richardson of the Rocke-
feller Foundation who provided the financial support
that made this study possible. It not only allowed
The Institute staff to have extensive contact with
representatives of state governments, but also en-
abled us to augment our staff with several outstanding graduate students.

We would like to sincerely thank Katherine Troll
and Joanne Polayes of the Yale School of Forestry and
Environmental Studies and Mary Jo Waits of the School
of Natural Resources at the University of Michigan.
Their contributions and efforts on behalf of this

project during the summer and fall of 1979 were key
to its success.

We also thank Judith Barnes for her contributions to the development and design of the mail
questionnaires.

Finally, we wish to thank four key colleagues
of The Institute staff. Judith Bayer, whose patience
in the typing of the numerous drafts of this manuscript is exceeded only by her outstanding capabilities as a professional; Jeanette Lendrum, for her
assistance in preparation of the original report;
William Hornick, whose assistance in editing the
paper and in producing the charts and graphics has
made the report a much improved and clearer statement; and Cecelia Rydell for the original cover art-
work.

Stuart L. Hart
Gordon A. Enk
Rensselaerville, N.Y.

Part I
The Analysis

1

Introduction

Since the signing of the National Environmental Policy Act (NEPA) on January 1, 1970, twenty-six states and one territory (Puerto Rico) have initiated state-level environmental review procedures. Of these twnety-six, fourteen states have established comprehensive legislation requiring the preparation of environmental impact statements (EIS's) on projects which may significantly affect the environment: California, Connecticut, Hawaii, Indiana, Maryland, Massachusetts, Minnesota, Montana, New York, North Carolina, South Dakota, Virginia, Washington, and Wisconsin. Michigan, New Jersey, and Utah have utilized executive orders to establish EIS requirements while Texas depends on administrative action to accomplish this end. *While executive orders carry the full force of the law, it is important to remember that they are imposed by the Governor--and can be withdrawn by the Governor. Similarly, requirements that are created by administrative action can be changed by it.*
New Mexico adopted EIS legislation in 1971 but repealed its law in 1972, after repeated attempts to weaken the EIS requirement. The remainder of the states (8) have established "limited" or "critical area" approaches requiring impact statements only for projects proposed within specific areas.
The State Environmental Policy Acts (SEPA's), or "NEPA equivalents," generally pattern themselves after the policies, goals, and even processes established by NEPA. They are "equivalent" in the sense that they utilize a similar tool--the environmental impact statement--and pursue a similar goal--the enhancement of environmental quality. The state laws, however, demonstrate a great deal of variability in terms of extent as well as style of implementation. In many cases, the variability includes

1

significant departure from the NEPA model. *Thus, innovation at the state level may indeed provide solutions to many of the shortcomings inherent in the original NEPA mandate.*[1]

This study focuses on the eighteen states which have, through legislation, executive order, or administrative action, initiated programs of comprehensive environmental review. Its purpose is to analyze how these states implement their impact statement requirement, to examine the costs associated with the EIS process, and to suggest ways of making the overall process more efficient and effective. To these ends, the report has been divided into two parts; Part 1 contains the comparative analysis of state programs as well as all conclusions and recommendations. Part 2 contains detailed supporting material pertaining to the evolution of each SEPA and its associated costs (this will be of most use to those interested in a detailed account of individual state programs). Finally, Appendix A--the SEPA "Fact Sheets"--provide a succinct "snapshot" of the current status of all eighteen SEPA's analyzed in this study.

DATA COLLECTION AND METHODOLOGY

Information on the state EIS programs was initially obtained through written requests sent to selected agencies within the eighteen SEPA states (Appendix D contains the names of individuals and agencies contacted). Given the time constraints of the study, the investigation was limited to the agency in each state responsible for *administering* the overall SEPA program, and the one or two most active state *project-originating* agencies. The first inquiry took place in mid-1976 and was continued through early 1977 as new sources of information (e.g., additional agencies) were identified.

The data collection task utilized a 21-page mail questionnaire designed so that agencies could respond to color-coded blocks of questions according to their role in the EIS process (see Appendix C). This included a request for official documents such as SEPA laws, guidelines, regulations, budgets, reports, or studies so that information provided by the respondent could be verified, balanced, or supplemented wherever appropriate.

The questionnaire information was further supplemented by means of in-depth, personal staff investigations in six of the eighteen states; Massachusetts, Minnesota, and North Carolina were visited

2

in early August of 1976, and <u>California</u>, <u>Washington</u>, <u>New York</u>, and <u>Massachusetts</u> (for a second time) were visited in November and December of 1977. The choice of states and agencies was based on several factors:

1. The six states cover the spectrum from the oldest state EIS program (California, 1970) to the newest (New York, 1975);
2. The states offer variety and contrast in SEPA implementation strategies and organization structures. In addition, the preliminary information indicated that the six states selected would be representative of a wide range of possible levels of SEPA implementation; and
3. All six states expressed enthusiasm and receptivity to being involved in the research.

Each of the eighteen states was contacted by telephone in January and February of 1978 both to finalize the information and to see if any significant changes had occurred since either the questionnaire response or the staff investigations. This proved very useful as several states had experienced important program changes in late 1977 or early 1978. Finally, applicable sections of the draft report were sent to each state in May of 1978 for review and comment so as to ensure factual accuracy.

Efforts to gather and assess the information on the various state EIS programs and their associated costs presented several difficulties. Due to the inherent variation among people and governmental programs, flexibility was essential in the research--especially in the area of costs. Since few states had actually appropriated money for their EIS programs and most agencies did not keep detailed records of EIS-related costs, it was necessary to approximate costs using other techniques. These included:

1. Using estimates of staff time expended, salary levels, and other approximations of expended costs (i.e., overhead, direct costs, etc.); and
2. Proportioning budget requests with actual appropriations.

It was not possible, given time constraints, to assess the costs of each environmental impact statement, and it was not considered desirable to estimate overall costs by using an "average" EIS figure

3

and multiplying it by caseload since the costs of
EIS's are so highly variable. Wherever additional
information was available (such as internally
generated program or cost studies), it was used to
supplement the components of costs assessed by our
inquiries. This was true in four states--California,
Minnesota, Washington, and Wisconsin--each of which
had conducted its own comprehensive assessment of
state EIS program statutes and costs.

Thus, we stress that the cost figures generated
by this study should be considered good *indicators
of magnitude* rather than precise documentation of
actual costs. Indeed, it is not possible (and may
not even be useful) to attempt the determination of
exact costs of the EIS process since a major goal of
all environmental impact statement programs is to
make environmental analysis a routine and integral
part of agency operations. As environmental considerations become better integrated into agency decision making, the costs associated with them become
harder to identify.

THE NATURE OF THE COSTS OF THE EIS PROCESS

Before embarking on our investigation of State
Environmental Policy Acts, it is important to understand the concept of the *full spectrum* of costs associated with environmental review although it may
never be possible to calculate these costs. To this
end, a matrix has been developed which provides a
graphic display of where these costs are incurred,
what form they take, and how they are interrelated[2]
(Table 1.1).

The matrix may be applied to an individual *project* or on a *program* basis at any level of government
by simply choosing the applicable participants in the
vertical column; these include:

- EIS Responsible Agency
- Project Originating Agency
- Review Agency
- Clearinghouse Agency
- Private Applicant
- Citizen/Consumer

Definitions are provided at the bottom of the matrix
and the spectrum of potential cost areas are arrayed
beneath each participant. Thus, for example, at the
state level, there may be separate "EIS Responsible"
and "Clearinghouse" agencies and several "Review"
agencies while, at the local level, all of these
functions might be performed by a single agency.

4

TABLE 1.1
EIS COST ACCOUNTING SYSTEM

COST ELEMENT / FUNCTION	Administration of the Law	EIS Preparation	EIS Review	EIS Distribution & Circulation	Delay			
					Inflation	Opportunity	Uncertainty	Mitigation
EIS RESPONSIBLE AGENCY[1]								
Salaries:								
Administration/Mgmt.[a]	+			+				
Review Staff			+					
Clerical Staff[b]	+		+	+				
Consultants	+		+					
Legal Counsel & Litigation	+							
Hearings (on Rules & Regs)	+							
Staff Training	+							
Public Info. & Assistance	+							
Overhead[c]	+		+	+				
Equipment, Materials & Telephone	+		+	+				
Office Space[d]	+		+	+				
Travel	+		+	+				
TOTAL DOLLAR COST	+		+	+				
PROJECT ORIGINATING AGENCY[2]								
Salaries:								
Administration/Mgmt.		+	+			+	+	+
EIS Preparation Staff		+				+	+	+
Incoming EIS Review Staff			+			+	+	+
Clerical Staff		+	+			+	+	+
Consultants		+				+	+	+
Legal Counsel & Litigation		+	+			+	+	+
Hearings (on projects)			+			+	+	+
Printing & Distribution				+		+	+	+
Overhead		+	+	+		+	+	+
Equipment, Materials & Telephone		+	+	+		+	+	+
Office Space		+	+	+		+	+	+
Travel		+	+	+		+	+	+
TOTAL DOLLAR COST		+	+	+		+	+	+
REVIEW AGENCY[3]								
Salaries:								
Administration/Mgmt.			+					+
EIS Review Staff			+					+
Clerical Staff			+					+
Consultants			+					+
Legal Counsel & Litigation			+					+
Overhead			+					+
Equipment, Materials & Telephone			+					+
Office Space			+					+
Travel			+					+
TOTAL DOLLAR COST			+					+
CLEARINGHOUSE AGENCY[4]								
Salaries:								
Administration/Mgmt.				+				+
Staff (EIS Collection & Distribution)				+				+
Clerical Staff				+				+
Overhead				+				+
Equipment, Materials & Telephone				+				+
Office Space				+				+
Travel				+				+
TOTAL DOLLAR COST				+				+

TABLE 1.1 (continued)
EIS COST ACCOUNTING SYSTEM

COST ELEMENT FUNCTION	Administration of the Law	EIS Preparation	EIS Review	EIS Distribution & Circulation	Inflation	Opportunity	Uncertainty	Mitigation
PRIVATE APPLICANT (if any) (prepare documents,								
Staff (supply information	+				+	+	+	+
Clerical Staff	+				±	+	+	±
Consultants	+				±	+	+	±
Legal Counsel & Litigation	+				±	+	+	±
Overhead	+				±	+	+	±
Equipment, Materials & Telephone	+				+	+	+	+
Office Space	+				±	+	+	±
Travel	+				±	+	+	±
TOTAL DOLLAR COST	+				±	+	+	±
CITIZEN/CONSUMER								
Donated Time		+						+
Consultants		+						+
Legal Counsel & Litigation		+						+
Equipment, Materials & Telephone		+						+
Travel		+						+
Added Cost of Goods and Services					±	+	+	±
TOTAL DOLLAR COST		+						±
SUMMARY								
Added Cost of Goods and Services (borne by Consumer)					±	+	+	±
Final Public Project Construction Cost					±	+	NA	±
Final Private Project Construction Cost					±	+	+	±

Notes

[1] The agency responsible for implementing the SEPA and overseeing other involved agencies. Term is used interchangeably with "Coordinating Agency."

[2] An agency that undertakes projects of the nature that would require an EIS as mandated by the SEPA.

[3] An agency that does not originate statements on projects with significant environmental effect but rather reviews EISs which are submitted by other agencies.

[4] The agency responsible for the review procedures of the SEPA by acting as a collection and distribution center for EISs.

[5] A private developer that needs government approval of some sort for a proposed project and may, therefore, be responsible for an EIS under SEPA.

[6] The EIS process frequently involves concerned citizens and citizen groups affected by a proposed project. In participating in the EIS review process, these citizens incur both direct expenses as well as the indirect but real expense represented by their expenditure of time. In addition, the consumer will pay higher prices for goods and services when the EIS process adds to project costs and these costs are passed along in higher prices.

[a] Managers and supervisors of those staff persons doing the actual EIS preparation, review and circulation work.

[b] Personnel assigned to record-keeping, typing, filing, etc.

[c] Includes both indirect expenses (i.e., heat, lights, etc.) and personnel benefits.

[d] All personnel assigned to EIS work require office space. This is a direct cost attributable to the EIS process, and applies to both public and private agencies.

The matrix arrays the cost elements of environmental review against the public and private bodies involved in the EIS process. Eight elements are displayed at the top of the matrix; the presence or absence of costs as well as their potential for being positive or negative factors is determined by following either a given function across the matrix or a given cost element down the matrix. Thus, for example, "inflation" may result in either a cost (-) or a savings (+) to a private applicant preparing an EIS in certain circumstances.

Although eight headings are shown at the top of the matrix, these may be conceptually arranged into the following four principal elements of cost:

1. Document Preparation, Review, Circulation, and Administration of the Law. This category (the first four headings on the matrix) includes the preparation cost of EIS's, negative declarations, and other environmental documents as well as the supporting tasks related to the distribution and review of these documents by appropriate agency personnel and members of the public. Administration of the Law pertains to the development of policies and principles for the SEPA statute. This would include, for example, the cost of promulgating rules and regulations, suggesting amendments to the statutes, training of staff, and informing the public as to what the law requires.

   These costs are incurred by both the private applicant and the public on *private projects* while *public projects* entail costs to the public sector only. Estimating these costs involves either obtaining budget and/or cost accounting data or assessing personnel time commitments (the costs of which often do not appear in any budget). *This category of cost constitutes the major focus of the cost analysis conducted in this study.*

2. Delay. The cost of delay attributable to the EIS process has two components:

   ● *Inflation* - increase in cost due to a time of generally rising prices for goods and factors of production. If construction is delayed during an in-

7

flationary period because of the EIS
process, the cost of that construction
may rise during the delay period. The
developer will thus pass along his increased costs to the consumer. If the
consumer's disposable personal income
does not rise as fast as the general
price level, then the increased cost
of the developer's product will represent a real added cost to the consumer.

- *Opportunity Cost* - the costs accruing
to an investor whose money might have
been used in alternative ways. For
both private and public projects, a
delay caused by the EIS process may
result in foregone net benefits or
revenues from a project.

In short, the cost of delay is a "complex,
non-linear function of time delay, type of
project, and overall economic factors such
as relative inflationary trends in various
segments of the economy."[3]

3. Uncertainty. The EIS process adds an element of administrative uncertainty to the
completion schedule of a proposed project
and this uncertainty has consequences in
terms of costs. Because the probability
of eventual project approval cannot be established quantitatively, the developer
often will require a higher rate of return
on his investment. The cost of abandoned
projects or projects never initiated because of uncertainty must also be recovered
or amortized and this applies to public as
well as private projects. These added
costs must be passed along to the consumer
in order to maintain the profitability of
the project.

4. Mitigation. The EIS process may result in
requiring the public or private developer
to moderate the impact of the proposed project. The process of achieving mitigation
often adds administrative as well as planning, engineering, and architectural expense. However, in order to estimate the
net costs associated with such environmental review, one must also consider the mone-

8

tary *benefits* associated with the mitigation of impacts. Mitigation (like inflation) may, therefore, increase or decrease the final cost of the project.

Thus, while the costs of *delay, uncertainty,* and *mitigation* are real and may be significant, they are exceedingly difficult to calculate and, for this reason, have been largely omitted from the cost analysis conducted in this study. Consequently, the analysis focuses on the costs of *program administration* and *document preparation, circulation,* and *review.* Calculating these costs, however, is no simple task either since, as was pointed out above, most agencies do not keep detailed records of EIS-related costs.

Acknowledging these difficulties should in no way detract from the value of the "EIS Cost Accounting System" (described above). It reflects more the limitations and weaknesses inherent in the administration of the environmental impact statement process.

With these caveats in mind, we turn to a comparative analysis of the eighteen State Environmental Policy Acts.

2

Key Issues in State-Level Environmental Review

In contemplating an environmental impact statement requirement a state is faced with several options concerning the scope of the program, who will
be responsible for it, and how it will be carried
out. The following addresses the "fundamental"
areas of the EIS process such as the statement of
policy, enforcement, and public participation. The
experience of the eighteen SEPA states in these
matters is comparatively analyzed and recommendations are offered based on findings related to these
experiences. More detailed descriptions of the individual state programs and their associated costs
may be found in Chapters 4 or 5 or in Appendix A--
the SEPA "Fact Sheets."

THE POLICY STATEMENT

The legislation enacted by most states follows
the NEPA statute quite closely. Indeed, most begin
with a broad declaration of environmental policy
similar to that of Section 101 of NEPA followed by
an "action-forcing" provision similar to the environmental impact statement requirement of Section 102
(2)(C) of NEPA. This historical trend in format is
particularly important *since the closer a state provision is to the language of NEPA and its guidelines,
the more likely it is that state courts will draw on
case law under NEPA and similar state acts in interpreting it.*[4]
Despite NEPA's model, however, many states have
developed variations on the traditional theme. Most
of these serve to make explicit what is only implicit
in NEPA. An interesting change of language appears
in the Maryland statute, which contains requirements
that all state agencies must "conduct their affairs
with an awareness that they are stewards of the air,

land, water, and living resources..." Similar language appears in the New York law, the most recently enacted EIS statute. Further, both statutes require that the state agencies "have an obligation to protect the environment for the use and enjoyment of this and all future generations." This language is not limited by the NEPA language "to use all practicable means" to protect resources; in addition, the introduction of the idea of "stewards" for the environment would appear to create an extremely strong state policy--almost the equivalent of a *direct right of citizens to a decent environment,* similar to constitutional amendments enacted along these lines in a number of states including Virginia and Montana.[5]

Virginia has taken this one step further by incorporating its constitutional mandate into a statutory statement of environmental policy *separate* from the environmental impact statement requirement. Hawaii has also chosen to enact statutes which separate the statement of environmental policy from the EIS requirement; South Dakota is the only state to have passed comprehensive EIS legislation while failing to provide for a statement of environmental policy either in the same law or by separate statute.

In addition to the setting of policy, state statutes are similar to (and different from) the Federal Act in other ways. For example, Washington also has the equivalents of 102(2)(A) and (B), i.e., the requirements for an interdisciplinary approach to environmental decision making and the development of methodology to ensure that "presently unquantified environmental amenities and values will be given appropriate consideration in decision making..." Similar provisions are found in the statutes of most other states.[6]

RECOMMENDATION #1

> *Each state should make a strong and clear policy statement concerning its environment. It is also important that this statement be geared to the specific needs and desires of the state (both in format and content); the extent to which the body of legal precedent established by NEPA applies depends on the degree to which states create innovative and original programs of comprehensive environmental review.*

On the one hand, it may be advisable to *separate* the statement of policy from any impact statement requirement since such an arrangement allows each to stand on its own merits, unencumbered by amendment or judicial interpretation of the other. On the other hand, a clear tie between a strong statement of policy (e.g., Virginia's and Montana's guarantee of citizens' *right* to a decent environment) and an EIS requirement may mutually reinforce one another. Perhaps a good example can be found in Maryland's law which states that the optimal balance between economic factors and environmental factors "requires the most thoughtful consideration of ecological, economic, developmental, recreational, historic, architectural, aesthetic, and other values." Such a statement recognizes the difficulty in mandating the development of specific tools, but at the same time presumably creates a requirement that environmental amenities be considered.

ELEMENTS OF THE ENVIRONMENTAL IMPACT STATEMENT

The National Environmental Policy Act requires five specific points of information to be covered in a Federal environmental impact statement:

1. the environmental impact of the proposed action;
2. any adverse environmental effects which cannot be avoided should the proposal be implemented;
3. alternatives to the proposed action;
4. the relationship between local short-term uses of man's environment and the maintenance and enhancement of long-term productivity; and
5. any irreversible and irretrievable commitments of resources which would be involved in the proposed action should it be implemented.

Guidelines for the preparation of Federal EIS's issued in 1973 by the Council on Environmental Quality (CEQ) added three more subjects which must be included in these statements:

1. a description of the proposed action and a statement of its purposes, along with a discussion of the environment affected;
2. an explanation of how the proposed action relates to land use plans, policies, and controls for the affected area; and
3. an indication of what other interests and considerations of Federal policy are thought to offset any adverse environmental effects outlined in the environmental impact statement.

13

Most states have adopted the five traditional provisions contained in the original NEPA mandate, although a few states chose to omit some requirements. Of greater significance, however, is the fact that *several states have added (or made more explicit) requirements which are only implicitly contained in the Federal law.*

Thus, California was the first state to add the statutory requirement that EIS's include a discussion of *mitigation measures* proposed to minimize the impact of development. Since then, several other states have adopted similar clauses requiring such discussion (see Table 2.1). Although NEPA itself does not specifically address this area, the NEPA guidelines do call for similar treatment of the subject in Federal EIS's.

TABLE 2.1
STATES WHICH MANDATE A DISCUSSION OF
"MITIGATIVE MEASURES" IN EIS's

	By Statute or Executive Order	By Regulation
California	X	
Connecticut	X	
Indiana		X
Massachusetts	X	
Michigan	X	
Montana		X
New Jersey		X
New York	X	
North Carolina	X	
South Dakota	X	
Utah	X	
Virginia	X	
Washington		X
Wisconsin		X

The California law was also the first to add the requirement that an environmental impact statement contain an analysis of the *growth-inducing impact* of a proposed action. While several other states require an EIS to include an evaluation of primary and secondary effects of a proposal besides California, only New York and South Dakota have added an explicit statutory requirement for consideration of growth-inducing impacts while Indiana and Montana address this issue through regulation. Hawaii's statement of environmental policy includes a unique provision for the setting of *population*

limits. Once again, although NEPA itself does not
specify this consideration, the NEPA guidelines place
strong emphasis on discussing the effects of a proposed project on population and growth.

In January of 1975, the California law was
amended to require that the mitigation measures' discussion include consideration of *energy conservation
measures* in order to reduce inefficient and unnecessary consumption of energy. Since then, Connecticut,
Texas, and New York have incorporated similar considerations into their EIS requirements and Indiana,
Minnesota, and Washington have issued regulations on
this point. Draft regulations for NEPA recently issued in response to a 1977 Executive Order also include a provision for discussing the energy requirements and conservation potential of the various alternative and mitigation measures.

A handful of states specify that the EIS include some discussion of the *economic impact* of proposed actions. Minnesota requires consideration of
any "direct and indirect adverse environmental, economic and employment effects" while the Wisconsin
law mandates discussion of the "economic advantages
and disadvantages" of a proposed project; Michigan
assesses only the costs involved in mitigation while
Washington requires a discussion of overall economic
impact through its regulations. The Federal guidelines are not as specific in this regard. They provide only that an EIS deal with "changed patterns of
social and economic activities" under the discussion
of secondary consequences of a proposal.[7]

Three states make explicit the requirement that
a *cost-benefit analysis* be performed. Connecticut
legislatively mandates that "an analysis of short-
term and long-term economic, social, and environmental costs and benefits" is to be conducted for
each proposed action. Montana and Washington require similar analyses by regulation. The problem
of cost-benefit analysis, however, is particularly
difficult. Although NEPA implicitly requires such
analysis, agencies have had difficulty in preparing
objective cost-benefit analyses and the courts have
had difficulty reviewing them.[8]

Maryland, Texas, and Wisconsin require that the
beneficial impacts of a proposal be discussed. However, Maryland is unique in that it also requires
that EIS's consider measures to "maximize potential
environmental effects." Again, while such ·language
is implicit in Federal law, it is not made explicit
nor is the "maximization of potential benefit" language made explicit by the CEQ guidelines or draft

15

regulations which mention only mitigation measures.

Three States' programs contain significant departures from the general model of EIS content. In Minnesota, an EIS must include an assessment of "the impact on State government of any Federal controls associated with proposed actions" and a discussion of "the multi-state responsibilities associated with proposed actions." Michigan's approach augments that of NEPA in that considerations of "human ecology" are specified in the EIS. Finally, Hawaii's implementing regulations specify that a list of "organizations and persons consulted" must be contained in every EIS.

Thus, it can be concluded, with few exceptions, that when the state acts and executive orders are taken in combination with their implementing guidelines, regulations, and procedures and compared with NEPA and its CEQ directives, there are only minor differences in their practical effect. Nonetheless, some specific recommendations can be made concerning the elements to be explicitly required in impact statements.

RECOMMENDATION #2

> States should strive, through legislation, to provide as wide ranging but explicit a mandate as possible for EIS content. Since the indirect effects of an action may be even more substantial than the primary effects of the action itself, clear consideration of the secondary and growth-inducing aspects associated with a project is of the utmost importance. It is also recommended that the EIS discuss the beneficial as well as adverse effects of a given project on natural, social, and economic conditions. The use of traditional cost-benefit analyses, however, is discouraged given their difficulty both in preparation and interpretation. A discussion of the mitigation measures to minimize the adverse effects should be explicitly required in all cases while consideration of the effects of a proposed action on the use and conservation of energy resources should be included where applicable. Finally, all alternatives should be required to undergo the spectrum of considerations outlined in the EIS, including those outside the jurisdiction and/or capability of the project proponent and the no-go alternative.

CRITERIA FOR EIS PREPARATION

The National Environmental Policy Act requires
an impact statement on "legislation and other *major*
Federal actions *significantly* affecting the quality
of the *human* environment." Although adhering fairly
closely to the *format* of the Federal criteria, states
have varied considerably in deciding whether their
EIS requirements should apply only to "major" actions or to all actions which may have a significant
effect; and whether only the physical environment
should be considered, or the totality of man's sur-
roundings, including physical, social, economic, cultural, and historic factors (i.e., the "human" environment).

Only Wisconsin has elected to use the verbatim
wording of NEPA although four other states (Minnesota,
New Jersey, South Dakota, Virginia, and Washington)
retain the requirement that an action be "major" before mandating EIS preparation. Indeed, Virginia
states in its legislation that only State construction projects costing $100,000 or more (excluding
highway projects) require impact statements. The
New Jersey Executive Order provides that all State-
funded or State-sponsored construction projects in
excess of $1 million or those projects less than $1
million but located in environmentally sensitive
areas be subject to environmental review. This follows the belief that large activities will be more
likely to have significant effects and that government will become bogged down if it has to prepare and
review too many documents. There may be, however,
many small-scale projects which, either singly or
cumulatively, have a significant effect on the environment. The remaining thirteen SEPA states have
generally followed the lead of California by specifying no generic size limitation in legislation,
thus making all actions with potentially significant
impact subject to some degree of environmental review.

The vast majority of state acts and executive
orders have also abandoned the NEPA language that
the "human environment" be impacted before an EIS becomes necessary; only Indiana, Montana, Texas, and
Wisconsin retain this wording. Michigan has adopted
the unique approach of requiring EIS's on actions
which "may have a significant impact on the environment or human life." Thus, Michigan's approach augments that of NEPA in that considerations of *human
ecology* are clearly articulated. Less explicitly,
Hawaii has broadened the wording of its SEPA to re-

17

quire EIS preparation on any action which will "probably have significant effects." Likewise, Massachusetts considers all actions with potential for "damage to the environment." Maryland, on the other hand, has defined its criteria as including actions "significantly affecting the quality of the environment, natural as well as socioeconomic and historic." Furthermore, if the impact is determined to be significant, *either adverse or beneficial*, an EIS is required under the Maryland Act.

RECOMMENDATION #3

It is recommended that the initial statement of criteria for EIS preparation be as comprehensive as possible, limited to neither a generic project size nor considerations of particular facets of the total environment.

Indeed, legislation or executive order is not the place for categorical exemption or criteria for threshold determination. These matters can be dealt with at great length in implementing guidelines or regulations. Thus, the example of Hawaii--where an EIS is required for any action which will "probably have significant effects"--may be worthy of study since it commits itself to no particular preconceptions. Conversely, the Michigan approach of recognizing the reality of human dominance in ecosystems may also be useful.

EXTENSIVENESS OF EIS APPLICABILITY

An important aspect of state EIS requirements concerns the applicability of those requirements to local governments and private development. All eighteen states which have general EIS procedures require impact statements for certain actions or projects undertaken directly by state agencies. There is great variation among the states, however, in their application of the EIS process to *local government actions* and to *private activity* for which a governmental permit is required (see Table 2.2).

Whether a state EIS requirement extends to local government and whether it includes private activities requiring public permits are critical issues, separate but overlapping. The private sector conducts a great range of activities that might have harmful environmental consequences, and many of them are subject to some form of discretionary governmental approval. The most important controls over pri-

18

vate actions, particularly those relating to the use
of land, are normally administered not by the states
but by counties, cities, villages, towns, or special
purpose units of local government. And apart from
their role of regulating private activities, local
governments are responsible for facilities and services intimately related to environmental quality--
for example, sewage and solid waste disposal, road
construction, water supply, and flood control. *Thus,
the real impact of a state's EIS program largely depends on whether it applies to local governments
and/or private activities.*[9]

TABLE 2.2
EXTENSIVENESS OF EIS APPLICABILITY
BY STATE

MANDATE APPLIES TO:
- Projects or actions <u>directly undertaken by</u>   *CN, IN,*
  <u>state agencies</u>; and   *MD, VA*
- Use of <u>state funds</u>.

MANDATE APPLIES TO:
- Projects or actions <u>directly undertaken by</u>   *MI, TX,*
  state agencies;   *UT, MN,*
- Use of <u>state funds</u>; and   *NJ, SD,*
- Actions requiring <u>state permits</u>.   *WI*

MANDATE APPLIES TO:
- Projects or actions <u>directly undertaken by</u>
  <u>state agencies</u>;
- Use of <u>state funds</u>;   *HI, MA,*
- Actions requiring <u>state permits</u>; and   *MN, NC*
- <u>Limited</u> projects or actions undertaken,
  funded or approved by <u>local agencies</u>.

MANDATE APPLIES TO:
- Projects or actions <u>directly undertaken by</u>   *CA, NY*
  <u>state and local agencies</u>;   *WA*
- Use of <u>state and local funds</u>; and
- Actions requiring <u>state and local permits</u>.

---

*Only three states comprehensively apply the EIS
requirement to both private activities and to local
governments.* <u>New York</u>, capitalizing on the experience in other states, is the first to have mandated
this degree of applicability in the original legis-
lation--before either legal interpretation or the
adoption of guidelines clarified the point. <u>Cali-</u>
<u>fornia's</u> law, through amendment is now also explicit
on both points. <u>Washington's</u> act, however, is limited to "all branches of government of this state, in-

cluding state agencies, municipal and public corporations, and counties" and thus depends on guidelines
to clarify its applicability to private activities
requiring state or local permits.

The implications of local-level implementation,
however, should in no way be assumed to be of equal
significance from one state to the next. Indeed, an
often overlooked aspect of this is the *political
structure* of a given state. California, for instance,
has only two levels of general-purpose local govern-
ment--counties and cities. About 80 percent of the
population lives within the boundaries of 400 cities
which range in size from that of the City of Los
Angeles (population 2.8 million) to those of small
communities of a few hundred. The fifty-eight counties are responsible for providing municipal services
to the unincorporated areas outside city limits.[10]
Thus, excluding special purpose districts, there
are only 458 units of local government affected
by California's EIS law. The political structure is
similar in Washington where, excluding special districts, there are only 306 units of local government
(39 counties and 267 cities).

New York, on the other hand, like much of the
eastern United States, utilizes two additional levels
of local government--townships and villages (townships being subdivisions of counties and villages
being small settlements within townships). As a result, some 1,612 units of local government (62 counties, 62 cities, 931 towns, and 557 villages) are
affected by the state's Environmental Quality Review
Act, excluding special districts. Furthermore, it is
the smaller-scale local governments (i.e., towns and
villages) that often bear the greatest burden since
they are usually the fastest growing areas (relative
to cities) and traditionally have the smallest bu-
reaucracies, the least resources, and the fewest personnel for effective incorporation of environmental
review. In this respect, New York's law is a unique
social experiment in the "grassroots" assessment of
environmental impact. *Thus, it is important to consider differences in political structure, primarily
between the eastern and western United States, when
analyzing the extensiveness of EIS applicability.*

Perhaps in partial recognition of the above,
four states have chosen to apply the EIS requirement
to localities only to a limited extent. In Hawaii,
the EIS requirement applies to actions involving "the
use of state or county lands or the use of state or
county funds" and to private actions within certain
designated areas on the state land-use plan. The

20

Massachusetts act applies to projects or activities initiated, financed or permitted by state agencies as well as the actions of local redevelopment authorities, housing authorities, and development commissions. Not included, however, are projects of local city or town agencies which do not need permits or licenses from state agencies and involve no state funds. The state of Minnesota is somewhat less specific by requiring that an EIS be prepared on any "major governmental action or for any major private action of more than local significance." Finally, while North Carolina requires EIS's only for actions involving the "expenditure of public moneys," it authorizes (but does not mandate) localities to require EIS's for major private development projects.

It is interesting to note that no states apply the EIS requirement to local government without also specifying that an impact statement be prepared on at least certain types of private activities subject to local public permission. A number of states have, however, followed the example of NEPA by requiring, either implicitly or explicitly, that an EIS be prepared on private activities for which a state (Federal in the case of NEPA) permit is required. These include Michigan, Montana, New Jersey, South Dakota, Texas, Utah, and Wisconsin.

Only three states--Indiana, Maryland, and Virginia--specifically stipulate that the EIS process should not be applied to the state permitting process. Indiana states unequivocally in its legislation that it "shall not be construed to require an environmental impact statement for the issuance of a license or permit by any agency of the state" and while the Maryland law only implies it, the guidelines specify that an impact statement is not required for the issuance of individual licenses or permits. The Virginia EIS requirement applies only to "major state projects." Only Connecticut remains somewhat vague with respect to this point; this may be clarified by regulations currently in the process of circulation and review.

RECOMMENDATION #4

It is strongly recommended that states require environmental impact statements for both public and private projects at the local as well as state level.

Since most development takes place at the local level through the granting of permits for private projects, the real impact of a state's EIS program depends largely upon whether it applies to local governments and private activities. Consideration must, however, be given to *political structure* and the *fiscal realities* associated with it. If environmental review is to be properly incorporated in the many small towns and villages of a state like New York, for example, a proper commitment to assist these localities (both monetarily and technically) must accompany the state's commitment to environmental quality.

ENFORCEMENT

All of the eighteen states which have comprehensive EIS requirements designate an agency to coordinate the program and to develop framework guidelines or rules and regulations for its implementation. That agency, however, usually has no specific authority to ensure that an impact statement is prepared (threshold determination) or to reject inadequately written statements (see Table 2.3).

This absence of clear enforcement authority follows the Federal example. Under NEPA, the Council on Environmental Quality was initially responsible only for issuing guidelines and assisting the various agencies in preparing their own procedures for implementing the Act. An executive order issued in 1977, however, empowered the CEQ to issue regulations binding on all Federal Agencies.

Despite this new rule-making authority, however, responsibility for compliance with the Act still rests with the agency proposing the action. NEPA specifies that copies of Federal impact statements must be filed with the CEQ, but does not say what is to be done with them; the Council has no veto power over agency proposals and cannot reject an inadequate statement.[11] Although CEQ's role can be described as largely advisory, the fact that decisions of environmental agencies are subject to review by the President provides the CEQ with the potential to influence actions at the highest level. Furthermore, Section 309 of the Clean Air Act[12] requires the Environmental Protection Agency (EPA) to review and comment on Federal agency proposals subject to the EIS requirement. If the Administrator of the EPA determines that any such proposal is unsatisfactory, he must make public his determination and refer the matter to CEQ.

Table 2.4 breaks the states down according to type of Coordinating Agency. Ten states have given responsibility for their EIS programs to environ-

TABLE 2.3
COORDINATING AGENCY AUTHORITY

Does the Coordinating Agency have the power to...

	Establish Statewide Guidelines or Rules and Regulations?	Review Agencies' Implementing Regulations?	Make Threshold Determinations?	Reject Inadequate EIS?
California	Yes	No	No	No[G]
Connecticut	Yes[P]	No	Yes[P]	Yes[G]
Hawaii	Yes[B]	N/A	No	Yes[G]
Indiana	Yes	Yes	No[C]	Yes
Maryland	Yes	No	No[C]	No
Massachusetts	Yes[B]	N/A	Yes	No
Michigan	Yes[I]	No	No	Yes
Minnesota	Yes	No	Yes[A]	Yes
Montana	Yes	No	No	No
New Jersey	Yes	No	Yes	Yes[G]
New York	Yes	Yes	No	No
North Carolina	Yes	No	No	Yes[G]
South Dakota	Yes[I]	No	No	No
Texas	Yes	No	No	No
Utah	Yes	No	Yes[G]	Yes[G]
Virginia	Yes	No	No	Yes[G]
Washington	Yes	Yes	No[C]	No
Wisconsin	Yes	No	No[C]	No

A = Only where original decision is appealed.
B = Binding on all agencies.
C = Criteria for decision supplied by the Coordinating Agency.
G = Power resides with the Governor's Office (or Chief Executive of local government).
I = Informally issued; not codified into administrative regulation.
P = Pending.

mental councils, boards, or interagency committees
similar to the Federal Council on Environmental Quality, while in five states this function is assigned
to a unit within the "natural resources" agency.
Only in three states is the duty split between the
natural resources department and the Governor's Office. Montana, however, is unique in that its Coordinating Agency--the Environmental Quality Council--
is located in the *legislative* rather than in the executive branch thus allowing it to operate with
greater effectiveness as a program watchdog.

TABLE 2.4 TYPE OF COORDINATING AGENCY			
Environmental Council, Board or Committee		Natural Resources Department	Nat. Res. & Governor's Office
Conn.	Montana	Massachusetts	California
Hawaii	Texas	New Jersey	Maryland
Indiana	Utah	New York	N. Dakota
Mich.	Virginia	South Dakota	
Minn.	Wisconsin	Washington	

     At present, all states have issued general
guidelines or rules and regulations for program implementation except Connecticut where regulations
are pending approval. Only three states--Indiana,
New York, and Washington--empower the Coordinating
Agency to review the regulations prepared by implementing agencies; most other states do, however,
mandate that individual agency programs be "no less
protective" of the environment than that contained
in the statewide guidelines. Notable exceptions to
this general rule are Hawaii and Massachusetts where
regulations prepared by the Coordinating Agencies are
binding on all affected agencies.
     We now focus more intently on the two most im-
portant aspects of enforcement--threshold determination and rejection of inadequate EIS's.

Threshold Determination

     The determination of whether or not an EIS is to
be required for a given action is, in many ways, the
key to the whole EIS process. Since an in-depth
analysis of environmental impacts cannot occur until
the potential significance of an action is recognized, the *procedure* as well as the *criteria* for making
such a decision are of the utmost importance.

24

Most states allow proposing or lead agencies to
determine whether or not impact statements are required for actions which they initiate or sponsor
according to their own criteria (see Table 2.5).
These criteria may or may not be subject to review
by the Coordinating Agency. Indiana, Maryland,
Washington, and Wisconsin, however, issue specific
criteria to agencies (in the form of a preliminary
"environmental assessment form" to be used in determining the need for further environmental review.

TABLE 2.5
THRESHOLD DETERMINATION

*Proposing or Lead Agency Determines Need
for EIS According to:*

their own criteria		*criteria developed by Coord. Agency*	*their own criteria but appeals are subject to review*	*Coordinating Agency Determines Need for EIS According to Their Own Criteria*
Calif.	N.Car.	Indiana	Minnesota	Connecticut
Hawaii	S.Dak.	Maryland		Massachusetts
Mich.	Texas	Washington		New Jersey
Mont.	Virg.	Wisconsin		Utah
N.Y.				

*Only in four states is the ultimate power of
threshold determination vested in a party separate
from the project sponsor or initiator.* In Connecticut, this decision is made informally (pending codification by forthcoming regulations) by the Department of Environmental Protection. In Massachusetts,
all project notifications and assessments must be
submitted to the Secretary of Environmental Affairs
for determination of the need for impact statement
preparation. The Office of Environmental Review within the New Jersey Department of Environmental Protection is charged with reviewing all preliminary assessments prepared under the Executive Order and determining, within thirty days, the need for the sponsor to prepare a full EIS; in Utah, the Governor retains the ultimate power to require the preparation
of EIS's.
Prior to amendments to the rules and regulations
in February of 1977, the Minnesota Environmental Quality Council also held the power of threshold determination. The Council had become so overburdened
with the hearing of individual proposals for EIS determinations, however, that it had little time to

address its other responsibilities. The amended regulations, therefore, passed this responsibility on to state agencies and local units of government; the EQC now makes the decision as to the need for an EIS only if valid objections are made to the sponsoring agency's decision.

RECOMMENDATION #5

> *State agencies and local governments should retain the responsibility of determining whether or not EIS's are required for projects which they sponsor or approve. Although threshold determinations should be routinely made by the agencies and localities, the ultimate decision should rest with neither the project sponsor nor the project originator. The Coordinating Agency, instead of the courts, should be the ultimate arbitrator in cases where the sponsoring agency's judgment is questioned.*

A third party (preferably the Coordinating Agency) should also be empowered to review the implementing guidelines and assessment criteria of individual agencies and localities for consistency with the overall state program. The issuance of *binding* criteria through regulation by the Coordinating Agency may work well at the state level but is not recommended where EIS requirements extend to local governments given the great variability of project type and scale from one level of government to the next.

Rejection of Inadequate EIS's

The absence of a "police power"--an agency empowered to reject inadequate EIS's--under NEPA is probably the major reason why so many impact statements have been found inadequate by the courts. Indeed, much of the potential for litigation could probably be eliminated if the various state acts provided for mechanisms allowing governmental agencies with particular expertise to force needed modification in another agency's statement prior to its adoption.[13]

In nine of the SEPA states, however, "enforcement" of the EIS process is limited to requiring a copy or notice of each impact statement to be furnished to the Coordinating Agency (see Table 2.6). Although many of these states allow for review of EIS's on a discretionary basis, none are empowered to do more than advise the initiating or sponsoring

agency as to acceptability. In Massachusetts, for
example, the Secretary of Environmental Affairs is
required to review every proposed project and "issue
a written statement indicating whether or not, in
his judgment, said reports adequately and properly
comply" with the law. The Secretary cannot, however,
order that an action be stopped on the basis of such
review.

TABLE 2.6
ENFORCEMENT AUTHORITY

I.   Coordinating Agency acts only as a repository for
     documents prepared.

     *California, Maryland, Massachusetts, Montana, New York,
     South Dakota, Texas, Washington, Wisconsin*

II. Coordinating Agency provides recommendations as to
     EIS acceptability to Governor's Office where ultimate
     rejection authority lies.

     *Connecticut, Hawaii, New Jersey, North Carolina,
     Utah, Virginia*

III. Coordinating Agency reviews EIS's for adequacy and
     may cause them to be resubmitted.

     Indiana, Michigan, Minnesota

With respect to enforcement, a handful of states
provide for some degree of discretion by higher authority. In Hawaii, North Carolina, Utah, and Virginia, the Governor retains the ultimate power to
reject inadequate EIS's. In Connecticut, the final
determination is made by the Office of Policy and
Management, an appendage to the Governor's Office;
in New Jersey, it is the responsibility of the State
Planning Task Force to reconcile all review conflicts
prior to clearing a project for construction.
    *Only in three states, however, is the Coordinat-
ing Agency given the power to reject inadequately
prepared EIS's.* In Indiana, the Environmental Management Board reviews the final EIS to ensure that
specified *procedures* were followed and that comments
received in the draft stage were properly addressed.
A decision on whether to accept the final EIS or require it to be revised is then made on this basis.
Thus, EMB has the power to reject inadequate EIS's
but depends heavily on the substantive comment of
reviewing agencies for its information. Conversely,

27

the Michigan Environmental Review Board, following a
technical review by an Interdepartmental Environmental Review Committee (INTERCOM), may determine that an
EIS is substantively inadequate and cause it to be
resubmitted. The Minnesota law, however, is definite-
ly the strongest. It authorizes the Environmental
Quality Council not only to require revision of inadequate EIS's, but also to *reverse or modify a proposed action if it determines that the action would
be inconsistent with the declaration of policy contained in the act.*
     At this juncture, it is important to remember
the profound effect which extent of EIS applicability
can have on the feasibility of enforcement. As an
EIS requirement is extended to include the permitted
actions of local government, it becomes more and more
difficult to provide an effective system of centralized review and enforcement. In California, for example, while state agencies prepared 200 EIS's in
1974, over 3,000 impact statements were prepared by
cities and counties. The situation was similar in
Washington where, since the law's inception in 1971,
110 EIS's have been produced by state agencies as
compared to 1,072 by local agencies. *Thus, while it
is legitimate to expect a state to enforce the EIS
requirement with respect to projects initiated or
sponsored by state agencies and even locally initiated or sponsored projects with regional or statewide
impacts, it is unreasonable and possibly counter-
productive to expect a state to review the thousands
of projects requiring EIS's strictly at the local
level.* It is probably for this reason that states
such as Indiana, Michigan, and even Minnesota are
able to provide for comprehensive enforcement while
California, Washington, and New York have been unable
(or unwilling) to do so.

RECOMMENDATION #6

> *All states should vest the Coordinating Agency
> with the power to reject inadequate impact
> statements prepared for projects initiated,
> sponsored, and, where applicable, approved by
> state agencies.*

In those states extending EIS requirements to local
governments, the Coordinating Agency should exercise
enforcement powers only over those locally initiated
or approved projects with regional or statewide implications. Given the vast number of privately initiated actions requiring local permits which are

potentially subject to EIS's, the routine determination as to adequacy of these documents should remain the prerogative of local government and should hinge on the granting of the required permits.

PUBLIC PARTICIPATION

Since agencies responsible for state environmental impact programs generally lack enforcement powers, the major burden of ensuring compliance falls upon citizen groups. By bringing deficiencies to the attention of administrators or by organizing political pressure, citizens have frequently been successful in persuading agencies to prepare or review EIS's. Agency officials are becoming increasingly responsive to the public because of the large number of court decisions supporting citizen demands and the ever-present threat of litigation.[14]
Only at the Federal level and in the "pioneer" states of California and Washington, however, have citizen suits been a major factor in either influencing or interpreting the EIS law. Even in these cases, the number of lawsuits has been miniscule when compared to the number of projects processed through the system. As the Commission on Federal Paperwork explains:

> ...CEQ estimates that between January 1, 1970 and June 30, 1975, 654 NEPA-related lawsuits were initiated, but that number should be seen in the context of the tens of thousands of Federal agency actions undertaken during that period. Thirty-three percent of the cases completed by June 30, 1975 were dismissed at the trial court level; 60 cases resulted in temporary injunctions (which halted projects for a few weeks or until an adequate EIS was prepared); permanent injunctions were issued in only four cases.[15]

In California, some 244 lawsuits were filed between 1970 and 1975 spanning the full range of issues encompassed under NEPA suits. These suits, however, affected less than 1 percent of all the projects for which EIR's and Negative Declarations were prepared during this period.[16] A similar experience is noted in Washington where between 1972 and 1976 in excess of 1,100 EIS's were prepared but only ten cases were heard by the State Supreme Court.[17]
Only eight of the other SEPA states have had more than a handful of cases involving their EIS

29

law.[18] Of these, the experience in Wisconsin has been most notable where the key litigation issue has shifted from EIS threshold decisions to the role of EIS's in regulatory decisions. The remaining eight states[19] have experienced no litigation whatsoever in connection with their SEPA. New York, however, can certainly expect to experience a certain level of litigation as its law becomes fully implemented in late 1978.

Although the ever-present threat of litigation has done much to sensitize the various agencies to citizen demands, there are many other less visible, but equally effective, means of public participation short of resorting to the courts (see Table 2.7). Like NEPA, however, most state laws and executive orders require only that copies of impact statements be made available to the public. In order to have any influence on an EIS (and ultimately how the EIS is used in decision making), citizens must, at a minimum:

1. know that a proposed project exists and that environmental documents have been prepared; and
2. have sufficient time to study and comment on the proposal.

How does the public find out about proposed projects and available environmental documents? At the Federal level, summaries of draft and final EIS's are published weekly in the Federal Register and monthly in CEQ's 102 Monitor. Twelve of the eighteen SEPA states issue a similar periodic, centralized list of EIS's and other documents.[20] This ranges from the simple listing of filed draft and final EIS's in the North Carolina Environmental Bulletin to the notice of permit applications, environmental assessments, draft and final EIS's, and Coordinating Agency meetings in Minnesota's EQC Monitor. *Thus, the level of public notice varies greatly even among those states with formal mechanisms of publication.* Another means of informing the public is through *notices in newspapers*. Ten states require agencies to give notice of draft EIS's in a newspaper of general circulation in the area affected. In Indiana, Montana, South Dakota, and Utah, however, EIS's are circulated to members of the public only to the extent deemed necessary by the originating or sponsoring agency.

What formal mechanisms have the states created to facilitate public participation? Most states require a minimum *public review period* of between

twenty and sixty days following public notice. Only
Wisconsin, however, requires a *public hearing* to be
held on every EIS. Minnesota stipulates that either
a hearing or a meeting is required as part of the
draft EIS review process; formal public hearings on
EIS's alone, however, are a rarity, primarily due to
cost. Informal public meetings are preferred because
it is not necessary to follow the formal hearing procedures such as cross-examination and preparation of
a hearing record.[21] In Connecticut, a public hearing is mandatory if twenty-five persons request such
a hearing within ten days of publication in the Connecticut Law Journal. Similarly, in Washington, a
hearing is required if fifty persons make written request; in Maryland, a meeting may be held through the
written request of fifty citizens. Most other states
"encourage" public hearings or informal meetings but
leave the decision to hold such meetings to the discretion of the initiating or sponsoring agency.

Nine states provide for the *formal participation
of the public in rule-making procedures*. In Hawaii,
for example, four months of public hearings on the
draft regulations provided opportunities for persons
on each Island to attend a hearing at a distance no
greater than forty-five miles from home. In California, Minnesota, New York, and Washington, the
testimony from public hearings as well as the review
of submitted comments were used in the drafting of
regulations.

While the Minnesota act makes no special allowance for citizen participation in reviewing impact
statements, it has experimented with two additional
mechanisms of citizen involvement. First, the law
contains the interesting provision that if a *petition*
signed by not less than 500 persons is filed with the
Environmental Quality Council, the Council "shall review the petition and, where there is material evidence of the need for an environmental review," require preparation of an EIS. Although the requirement of 500 signatures for a petition may make it extremely difficult for a single aggrieved individual
without organizational ties to present his grievance
to the Council, this mechanism does represent a
unique attempt to facilitate public involvement outside of the courts. Second, there is a statutory
provision for a Citizens' Advisory Council (CAC) to
advise the Environmental Quality Council, with the
basic objective of ensuring citizen participation in
the activities and decisions of the Council. All
eleven members of the CAC are appointed by the Governor with each of eight members representing a

TABLE 2.7
PUBLIC INVOLVEMENT MECHANISMS

State	Rulemaking Public Hearings	Citizen Petition	Public Notice	Scoping of Content	Public Hearings or meetings on Draft EIS	Citizens Advisory Committee	Citizens Suits
CT	No	No	Newspaper; CT. Law Journal	No	Mandatory if 25 persons request	No	No
CA	Yes	No	Newspaper; CA EIR Monitor	No	Discretionary	No	Yes
HI	Yes	No	Newspaper; EQC Bulletin	No	Discretionary	No	Yes
IN	Yes	No	EIS's circulated as deemed necessary	No	Discretionary	No	Yes
MD	No	No	Newspaper; quarterly lists	No	A meeting may be held through written request of 50 citizens	No	Yes
MA	Yes	No	Newspaper; Environmental Monitor	Limited access to scoping procedure	Discretionary	No	Yes
MI	Yes	No	EIS Status List	No	Discretionary	No	No
MN	Yes	Yes	Newspaper; EQC Monitor	No	Mandatory hearings or meetings	Yes	Yes
MT	No	No	EIS's circulated as deemed necessary	No	Discretionary	No	Yes

TABLE 2.7 (continued)
PUBLIC INVOLVEMENT MECHANISMS

State	Rulemaking Public Hearings	Citizen Petition	Public Notice	Scoping of Content	Public Hearings or meetings on Draft EIS	Citizens Advisory Committee	Citizens Suits
NJ	No	No	Weekly Bulletin	No	Discretionary	No	No
NY	Yes	No	Newspaper; Environmental Notice Bulletin	Yes	Discretionary	No	No
NC	No	No	Newspaper; NC Environmental Bulletin	No	Discretionary	No	Yes (standing uncertain)
SD	No	No	EIS's circulated to "interested" parties	No	Discretionary	No	No
TX	No	No	None	No	Discretionary	No	No
UT	No	No	EIS's circulated to "interested" parties	No	Discretionary	No	No
VA	No	No	Monthly listings	No	Discretionary	No	No
WA	Yes	No	Newspaper; SEPA Register	No	Hearing required if 50 persons make written request	No	Yes
WI	Yes	No	Newspaper	No	Mandatory Hearing	No	Yes

Congressional district and three members serving at
large. Unfortunately, the CAC, though an intrigu-
ing concept, has not functioned as a truly effective means of assuring citizen input. From the outset, it has been a compromise measure, representing
an attempt to provide the Council with some form of
institutionalized citizen input without going all the
way to establishing a citizens' decision-making board.

Only two states--Massachusetts and New York--
have incorporated provisions for *pre-draft scoping*
of EIS content; this will be discussed in greater detail in the next Chapter. Although the scoping process is mandatory in Massachusetts, and the Secretary of Environmental Affairs is responsible for determining both the scope and content of individual
EIS's, the extent to which the public should be included in the process is unclear. New York's scoping process, on the other hand, is optional, but
specifically states that "lead agencies shall make
every practicable effort to involve applicants, other
agencies and the public in the SEQR process."

In the final analysis, the greatest barrier to
providing adequate opportunity for public participation in the EIS process is in defining what constitutes "the public." Indeed, it is difficult to
identify the "general" public when dealing with individual project proposals of specific impact. As
a result, a few states have attempted to come up
with working definitions of the term. Hawaii, for
example, specifies that EIS's shall be disseminated
to "the various concerned agencies and citizens'
groups" while Maryland calls for comment by agencies
and "private organizations and individuals with jurisdiction by law, special expertise, or recognized
interest."

RECOMMENDATION #7

> *Given that public access to and participation*
> *in environmental decision making is a basic*
> *goal of the EIS process, states should aggres-*
> *sively solicit citizen participation through*
> *extensive public notice and a wide dissemina-*
> *tion of environmental documents.*

It is particularly important to get the public involved early in the process so that the *impact statement* reflects their concerns rather than relying on
litigation to force adequate consideration of the
concerned citizenry. This means that notices of application and preliminary assessments should be

widely published rather than waiting for the draft
EIS to be released before involving the public.
Since a flawless public involvement method has yet
to be developed, it is also important that states
provide a *variety* of avenues for public participation
(e.g., public hearings,informational meetings, citizen petition, scoping, etc.) since, in combination,
these techniques would tend to counter each other's
weaknesses.

From the foregoing, it should be clear that the
level of commitment and sophistication is highly variable among the eighteen SEPA states. Factors such
as *extensiveness of EIS applicability, enforcement,*
and *public participation,* while providing the under-
pinnings of an EIS program, however, also have direct implications in terms of economic *cost.* The
next Chapter begins with an analysis of environmental
review costs and closes with a discussion of adaptive
program techniques which can help to mitigate such
costs.

3

The Costs of State-Level Environmental Review

The previous Chapter demonstrated the spectrum of approaches and techniques currently being employed by the states in their efforts to implement programs of comprehensive environmental review. At the one extreme, for example, we have a state such as <u>Indiana</u> which exerts a minimum of effort on behalf of its state EIS program and presumably reaps few benefits beyond those provided through NEPA channels. At the other extreme, we have a state such as <u>California</u> which has expended a great deal of time and effort developing a strong, multi-level impact statement program but has also become acutely aware that wasteful or overlapping requirements do nothing except drain a program of its vitality. *In short, experience with SEPA's has raised the possibility of greatly improving decision making and planning through increased information and accountability but has also presented the spectre of overburdening state and local governments already hard-pressed to carry out vital planning functions as well as adding significant cost to projects in both the public and private sectors.*

Indeed, an often cited criticism of environmental programs in general and EIS requirements in particular is that such requirements encourage excessive government spending, impede economic growth, and contribute to an already burdensome problem of unemployment. This Chapter deals largely with the public sector costs of environmental review at the state level but also addresses, to a lesser extent, the other issues. All cost information is drawn from the extensive body of data gathered and recorded, as detailed accounts of each state program, in Chapters 4 and 5. After a general analysis of state EIS program costs, a number of techniques being used in the states are identified which have the effect

37

of mitigating unnecessary delay, duplication, and
overlap in the EIS process. Comparative analysis
of these approaches is followed by recommendations
intended to foster a more efficient and effective
EIS process.

PROBLEMS IN ASSESSING EIS-RELATED COST

As stated previously, it is not possible and
may not even be useful to attempt the determination
of *exact* costs of the various state programs of
environmental review. Like NEPA, a major goal of all
state-level requirements is to make environmental
analysis a routine and integral part of agency operations just as economic and technical analyses are.
And as environmental considerations become better
integrated into agency decision making, the costs
associated with them become harder to identify.

Furthermore, *most agencies do not keep detailed
records of EIS-related costs.* For example, EIS preparation is so often intermixed with other related
activities, including preliminary design, site reviews, permit application, and policy development,
that determining separable costs may not be possible.

This problem is compounded by the fact that *few
states have appropriated meaningful sums of money for
the implementation of EIS requirements.* Agencies are
often required to initiate environmental review within existing resources by shifting operating budgets
and personnel from other funded programs. As a result, many states, including California, Washington,
New York, North Carolina, and Wisconsin have suffered from inadequate staff resources to comply fully
with the law. The need for financial assistance is
especially critical in those states requiring local
jurisdictions to adopt programs of environmental review. Nowhere is this problem more evident than in
New York where statewide, only $132,791 has been ap-
propriated since 1976 for the implementation and administration of a SEPA law encompassing over 1,600
units of local government. The result has been the
extensive "borrowing" of already existing staff in
order to fulfill the minimum responsibilities under
the Act.[22]

*Despite a widespread lack of appropriations
among the states, the costs of minimum compliance
still exist and must be internalized within existing
agency programs.* The identification of this "borrowed" time is, therefore, critical to the meaningful estimation of EIS-related costs.[23]

THE COSTS OF ENVIRONMENTAL REVIEW

Coordinating Costs

Of all aspects of environmental review, the costs of administering and coordinating the program are the easiest to estimate since they are usually incurred by a central core of staff. States vary considerably in the amount of effort invested by the Coordinating Agency in such functions as guidelines preparation, program assistance, and document review. *That effort, whether funded by appropriation or through "borrowed" time, is generally a good indication of the overall level of program implementation.* As Table 3.1 shows, it is also indicative of the extensiveness of EIS applicability within the various states.

In Connecticut and Maryland, the coordination role is filled by the partial and unbudgeted commitment of a single person's time. The situation is much the same in Indiana and Virginia except that positions have been officially created to serve these roles. In Indiana, all twelve EIS's prepared thus far in compliance with SEPA have also involved a Federal EIS. Thus, the incremental costs are minimal.

In Michigan, Montana, and New Jersey, SEPA coordination responsibilities are a funded part of a larger board, council, or interagency body. In Michigan and New Jersey, this function includes the power to reject inadequate EIS's. Interagency bodies in Utah and Wisconsin also combine SEPA responsibilities with other functions; because of difficulty in separating out SEPA costs, the figures presented constitute *overall* agency appropriations. Since no full EIS's have as yet been prepared solely under the Utah executive order, however, the costs associated with that state program are probably minimal. The situation is similar in South Dakota and Texas where the EIS requirement has been little used and most comprehensive environmental review has also involved NEPA compliance.

The coordination costs in Hawaii, Massachusetts, and Minnesota all reflect the sums appropriated to a central Coordinating Agency which also reviews and comments on every EIS. In the case of Minnesota, this includes the power to cause inadequate EIS's to be resubmitted; Massachusetts and Hawaii are granted only indirect review authority either via the Governor (Hawaii) or the Attorney General (Massachusetts). In North Carolina, however, there has been little

```
┌─────────────────────────────────────────────────────────┐
│                      TABLE 3.1                           │
│           COSTS OF COORDINATION vs. EXTENSIVENESS         │
│                  OF EIS APPLICABILITY                    │
│                                                          │
│ MANDATE APPLIES TO:                                      │
│   ● Projects or actions directly undertaken by state agen-│
│     cies; and                                            │
│   ● Use of state funds.                                  │
│   Connecticut - $18,000-$36,000/yr.  Virginia - $19,000/yr.│
│   Indiana     - $27,000/yr.          Maryland - $7,000-  │
│                                                 $10,000/yr.│
│                                                          │
│ MANDATE APPLIES TO:                                      │
│   ● Projects or actions directly undertaken by state     │
│     agencies;                                            │
│   ● Use of state funds; and                              │
│   ● Actions requiring state permits.                     │
│   Michigan    - $60,000/yr.    South Dakota - negligible │
│   Montana     - $28,000/yr.    Texas        - negligible │
│   New Jersey  - $30,000/yr.    Utah         - $100,000/yr.│
│                                Wisconsin    - $300,000/yr.│
│                                                          │
│ MANDATE APPLIES TO:                                      │
│   ● Projects or actions directly undertaken by state     │
│     agencies;                                            │
│   ● Use of state funds;                                  │
│   ● Actions requiring state permits; and                 │
│   ● Limited projects or actions undertaken, funded, or   │
│     approved by local agencies.                          │
│   Hawaii        - $120,000/yr. Minnesota      - $91,285/yr.│
│   Massachusetts - $109,000/yr. North Carolina - $15,000/yr.│
│                                                          │
│ MANDATE APPLIES TO:                                      │
│   ● Projects or actions directly undertaken by state and │
│     local agencies;                                      │
│   ● Use of state and local funds; and                    │
│   ● Actions requiring state and local permits.           │
│   California - $178,400/yr.        New York - $316,000/yr.│
│   Washington - $150,000/yr.                              │
└─────────────────────────────────────────────────────────┘
```

effort to implement the requirement save two persons' "borrowed" time within the Coordinating Agency.

In California, Washington, and New York, while
appropriation has been made for program implementation and coordination, the amount appears to fall far
short of what is needed to properly institute such
comprehensive and diversified undertakings. A prime
example is New York where, since 1976, only $132,791
have been directly appropriated for coordinating the
State Environmental Quality Review Act; this compare
to "actual" costs which may run as high as $700,000

annually.[24] There is also growing concern in these three states over the problem of imposing new requirements on local jurisdictions without provision for financing the added costs involved.

Preparation and Review Costs

Unlike program coordination, the costs of document preparation, circulation, and review are extremely difficult to assess given the *integrative* nature of the EIS process and the *decentralized* nature of agency and public review. The only general statement that can be made concerning preparation and review costs is that, with few exceptions, states provide little if any additional money for the execution of these very tangible and sometimes expensive requirements. That is, most agencies charged with the preparation or review of EIS's receive no additional staff or funding to perform these functions; they are, in effect, forced to "borrow" from existing personnel and shift resources from other programs.

The magnitude of these costs varies greatly, not only from state to state, but also from agency to agency and project to project. In Connecticut, for example, preparation costs have ranged from $10,900 to $65,000 in the four EIS's prepared thus far under CEPA. The time spent in preparing EIS's in Montana has ranged from five people working three weeks (about $4,000) to twenty-five people working two years (about $675,000); agency staff hours expended in reviewing these documents have varied from less than one person-hour to perhaps sixty to eighty. The cost of document preparation in New Jersey is also highly variable given the diverse nature of project proposals. For example, an initial assessment ending in a Negative Declaration generally costs between $300 and $500 while a single EIS has cost as much as $200,000. Similarly, the costs of EIS preparation in Wisconsin have varied greatly and range from $6,000 to more than $150,000. In contrast, the primary objective of the Virginia EIS program has been to establish project compatibility with review agency goals early on. As a result, the cost of documentation has been minimal; on the average, "preliminary" EIS's are only five to six pages long and take only about eight hours to prepare.

The *mechanics of the EIS process* and the *criteria for document preparation* also have a great influence on the general level of preparation costs. In Massachusetts, for example, preliminary environmental

41

assessments require approximately one-tenth the
amount of work required on a full EIS since the former consists of only a standard four-page form. Prior
to the adoption of a similar preliminary "worksheet"
format in Minnesota, the assessments consisted of
lengthy essays comparable to full EIS's prepared
under some state programs (e.g., North Carolina where
a common length for an EIS has been only ten pages).

Several states such as Minnesota, Virginia, and
New Jersey have instituted specific threshold limits
defining what constitutes a project requiring an EIS.
In this regard, Virginia and New Jersey set monetary
thresholds of $100,000 and $1 million respectively.
Similarly, in Minnesota, only "major governmental
actions" or a "major private action of more than local significance" require EIS's. As a result, "units
of impact" are generally in the hundreds of acres,
thousands of housing units, or tens of miles with an
average project cost of nearly $200 million. However,
in a state such as Massachusetts, where all actions
with the potential for "damage to the environment"
require EIS's, the projects are generally of a more
medium scale with "units of impact" in the tens of
acres, hundreds of housing units, and single miles
with an average project cost of less than $20 mil-
lion.[25]   Thus, while more projects may come under review in Massachusetts, the average *scale* of these projects is an order of magnitude less than in a state
such as Minnesota as a result of defined threshold
limits.

Total Costs

Despite the problems inherent in assessing EIS-
related costs, four states--California, Washington,
Wisconsin, and Minnesota--have, nevertheless, attempted to "get a handle" on the total cost associated with their environmental review programs. The
matrix presented in Chapter 1 should again be consulted for a graphic display of where these costs are
incurred and how they interact.

Since any effort of this sort requires a number
of "leaps of faith" and is necessarily based on a
sample of the total rather than on an exhaustive examination of every agency and every project, the figures should be considered as good *indicators of cost
magnitude* rather than precise documentation of actual
costs. The text for each of the four states in Chapter 4 should be individually consulted for a more
thorough description of each cost study's constraints
and limitations.

The following excerpts, in combination with Table 3.2, provide a comparative capsule of the cost estimates generated by the statewide studies:

TABLE 3.2
SUMMARY OF STATEWIDE EIS COST
STUDIES IN FOUR STATES

	Extensiveness of EIS Applicability	Caseload EA's/EIS's	Total Cost
Wisconsin	State	1,573/65	$605,584 (FY 1975)
Minnesota	State Major private	221/60	$2.5 million $1974-1977)
Washington	State All local	State-19 EISs Local-168 EISs	$4.5 million (per yr. est.)
California	State All local	State-200 EISs Local-3400 EISs (1974)	State-$ 7 mil Local-$16.8 mil Priv.-$10 mil Total-$33.8 mil (per yr. est.)

- In Wisconsin, where the EIS requirement applies only to actions directly undertaken, supported, or approved by state agencies, the estimated total cost of WEPA-related activity was only $605,584 for the period July 1, 1974 to June 30, 1975. This entailed the preparation and review of 1,573 screening documents and 65 preliminary environmental reviews (similar to draft EIS's) or final EIS's.

- In Minnesota, where the EIS requirement applies to major actions directly undertaken, supported, or approved by state agencies as well as major private actions of more than local significance, the total cost of the program from 1974 to December 1, 1977 was estimated to be only $2.5 million, $330,000 of which was incurred by local governments. *This entailed the preparation and review of 221 preliminary assessments (75 percent prepared by localities) and 60 EIS's (13 prepared by locclities).*

43

- In <u>Washington</u>, where the EIS requirement applies to all actions undertaken, sponsored, or approved by state and local agencies, approximately 80 percent of the environmental documentation takes place at the local level (in 1976, 168 EIS's were prepared by localities as compared to 19 for state agencies). The total estimated cost for this program is approximately $4.5 million per year, $1-2 million of this being incurred at the local level. *Thus, while 80 percent of the EIS's are prepared at the local level, less than one-third of the cost is incurred there since state projects tend to be much larger in scale.*

- In <u>California</u>, where the EIS requirement also applies to all actions undertaken, sponsored, or approved by state and local agencies, approximately 90 percent of the environmental documentation takes place at the local level (in 1974, 3,400 EIS's were prepared by localities as compared to only 200 by state agencies). The total estimated cost of this program is broken down as follows:

State Agencies	$ 7 million
Local Government	$16.8 million
Private Applicant	$10 million
TOTAL COST PER YEAR	$33.8 million

When allowances for project delay, uncertainty, and mitigation costs are included, the cost is estimated to be in the range of $50-75 million per year. Even including these rather large allowances for economic conditions, however, this represents only about *0.5 percent of total project costs or $2-$3 per capita annually.*

Trends in Cost and Caseload

The State Environmental Policy Acts assumed the existence of a highly structured, well-coordinated, comprehensive decision-making system into which environmental factors could be incorporated. The lack of such organized decision pathways in many of the states, however, resulted in project backlogs and necessitated considerable *initial expenditures* aimed at reforming governmental practices so as to foster the coordinated review of environmental impact.[26]

In addition to these costs of initial "restructuring," there was also the one-time expense of establishing (often through the courts) an appropriate policy of *retroactive applicability* (i.e., which existing projects should, by virtue of their potential impact, be required to comply with the EIS law despite the fact that they were proposed prior to its passage). Indeed, one of the major weaknesses of the early California and Washington programs was their failure to address adequately this "grandfathering" issue.[27] Having the luxury of hindsight, other states have been able to deal more effectively with this problem. This is especially true in New York where recent revisions to the rules and regulations dealt extensively with the identification and exclusion of "grandfathered" projects from EIS requirements.

Thus, in implementing EIS requirements, states experience an "initial" period of *program establishment* followed by a "transition" period of *process clarification* and *backlog processing* (see Figure 3.1). As states gain more experience with their EIS programs, they slowly emerge from the "transition" period. This "maturing" process seems to be characterized by two trends in caseload--both of which have the indirect effect of reducing program costs:

- *The overall number of actions requiring EIS's appears to stabilize at a lower level than that experienced during the "transition" period; and*

- *Fewer EIS's are prepared relative to the number of Negative Declarations. The "crossover" in caseload occurs during the "transition" period and generally stabilizes at a level of about ten to twenty Negative Declarations for every EIS prepared.*

The former trend is prominent in several states although it is most apparent in the earlier or "pioneer" SEPA states where limited experience resulted in problems (such as the "grandfathering" issue) that were tactfully avoided by later state EIS programs. In Michigan, for example, while thirty-five EIS's and thirty-four Negative Declarations were acted upon in 1975, this number dropped to fifteen EIS's and fifteen Negative Declarations by 1977. Texas showed a similar decline; in 1975, twenty-six EIS's and forty Negative Declarations were prepared as compared to only four impact statements and seven

FIGURE 3.1
GENERALIZED DYNAMIC OF CASELOAD

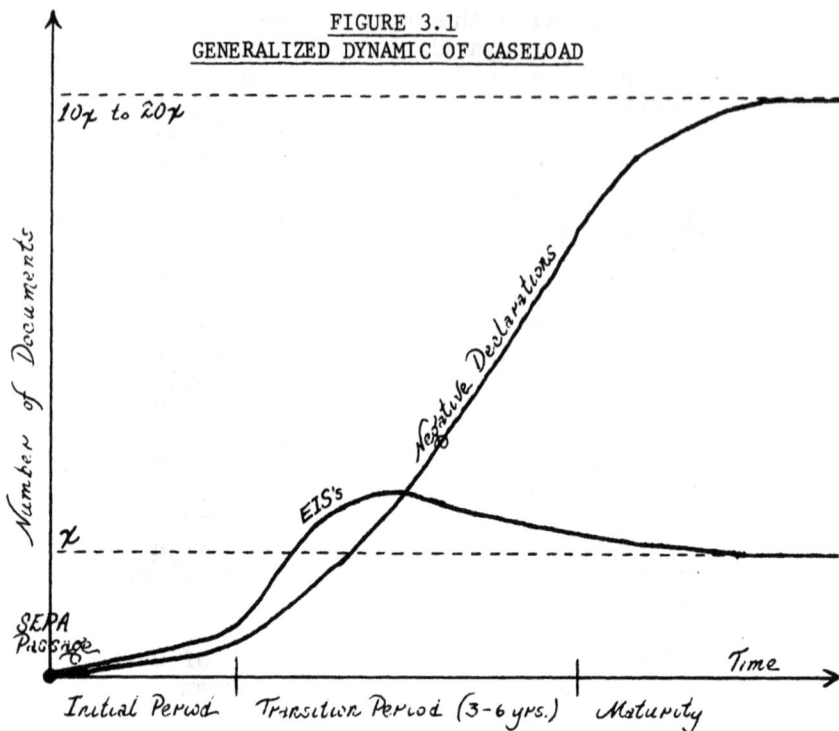

FIGURE 3.1
GENERALIZED DYNAMIC OF CASELOAD

negative findings in 1977. Massachusetts, however,
shows perhaps the greatest attenuation of environmental documentation. Consider the following figures:

MASSACHUSETTS CASELOAD

	FY 1975	FY 1976	FY 1977
Environmental Assessments	961	385	309
DRAFT EIR's	35	24	15

The situation in Montana, however, is probably
more characteristic of the actual dynamic in caseload; while thirty-one EIS's were prepared in fiscal
1974, this number dropped to ten in fiscal 1975 and
rose once again to nineteen in fiscal 1976 before
dropping to the current level of nine EIS's for fiscal 1977. Thus, while overall, the number of EIS's
prepared is declining, the process is anything but
smooth. The erratic nature of this decline may be
explained by the "feast-famine" syndrome which results in inflated caseloads for a short period of

time. In Montana, for instance, processing of major
applications for industrial water in the Yellowstone
Basin will begin when the Yellowstone Moratorium ex-
pires in 1978. This may mean up to ten additional
EIS's on top of the normal caseload to be completed
in 1979.[28]
    With regard to the second trend--the changing
ratio of EIS's to Negative Declarations--this phenomenon is also most apparent in the "pioneer"
states, particularly California and Washington. A
brief look at how the mix of environmental documentation in Washington has changed over the last five
years yields many interesting findings (see Table
3.3). Although the figures for the "draft EIS's
column include both SEPA and NEPA statements, it is
evident that the ratio between EIS's and Negative
Declarations is rapidly shifting in the direction
of the latter.

TABLE 3.3 FIVE YEARS OF ENVIRONMENTAL REPORTING IN WASHINGTON		
Year	Draft EIS's (SEPA + NEPA)	Negative Declarations
1972	175	24
1973	310	333
1974	384	487
1975	250	262
1976	217	369
5-Year Total	1,336	1,475

The trend is equally apparent in California:

TABLE 3.4 ENVIRONMENTAL DOCUMENTATION IN CALIFORNIA		
Year	State EIS's/Neg.Dec.	Local EIS's/Neg.Dec.
1975-76	208/151	819/1,599
1976-77	203/298	725/2,284

    At least partly responsible for this dramatic
change in the ratio of EIS's to Negative Declarations is the current emphasis placed on the preparation of *programmatic or generic impact statements*

at both the state and Federal levels. Widespread preparation of these program level analyses has greatly reduced the number of individual EIS's required. The phenomenon may, however, also be at least partially attributable to the gradual attainment, by agencies and localities, of *procedural compliance with the law*. Since nearly all interpretation of NEPA and its various state equivalents has come from the courts, and since nearly all legal decisions related to EIS's have been limited to procedural issues, the efficacy of litigation is impaired by such conformance. That is, by faithfully complying with all the <u>procedural</u> requirements of the law, agencies and project-originators can effectively "beat the system" by proceeding with their projects as planned. In fact, as we have seen, only two states--<u>Michigan</u> and <u>Minnesota</u>--specifically empower a governmental body to reject an EIS on the basis of substantive inadequacy. The rest, like NEPA, require only varying degrees of procedural compliance.

Indeed, even the prevailing cause of litigation has changed from complaints involving the *failure to file an EIS* in 1971 and 1972, to tests of the *adequacy of EIS's* prepared in 1975.[29] The Negative Declaration with attached mitigative requirements, therefore, becomes the most practical technique for achieving the <u>substantive</u> ends of the law. *Thus, the trend is really toward greater use of the Negative Declaration to achieve mitigation of impacts and away from litigation to achieve delay or injunction.* This is borne out by the fact that, aside from <u>California</u> and <u>Washington</u>, no other state has experienced more than a handful of court cases under an EIS law.

Conclusions

    1. *The first conclusion based on the information generated by this study is that, generally, most actions or decisions do not require the preparation of environmental impact statements (See Figure 3.2).*

Ignoring the estimated 80-90 percent of all actions that can be categorically exempted[30] from the EIS process (e.g., maintenance and repair activities, minor new construction, routine permitting, etc.), the vast majority of actions with the potential for adverse environmental effects require only a *preliminary environmental assessment* to document the non-

48

FIGURE 3.2
GENERALIZED BREAKDOWN OF ACTIONS
SUBJECT TO EIS PREPARATION

"All agencies shall prepare, or cause to be prepared, an EIS on any actions they propose, fund or approve which may have a significant effect on the environment."

Of all ACTIONS,

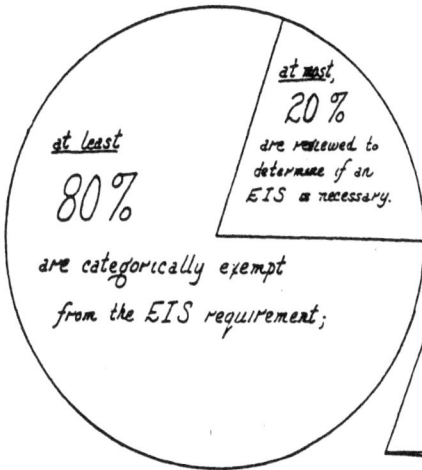

Of this 20% reviewed,

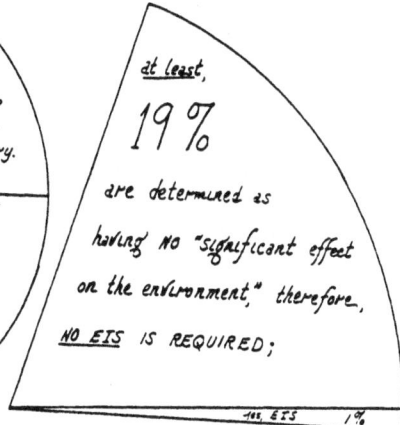

at least
80%
are categorically exempt from the EIS requirement;

at most,
20%
are reviewed to determine if an EIS is necessary.

at least,
19%
are determined as having No "significant effect on the environment," therefore, NO EIS IS REQUIRED;

at most, 1% are determined to need an EIS because of potential effects.

Of this 1% on which an EIS is required,

at least, 4/5% receive routine approval;

at most, 1/5% produce delays, litigation, abandonment or other complications...

in other words, only ONE in every FIVE HUNDRED.

49

significance of their impact. A number of examples
may be cited:

State/Year	Negative Declarations	EIS's	Ratio
Hawaii/1977	⌣350	33	10:1
Maryland/1976	156	20	8:1
Massachusetts/1977	309	15	20:1
New Jersey/1977	28	2	14:1
New York/Total	399	25	16:1
Montana/1976	193	10	20:1
Utah/1976	25	0	-
Wisconsin/1975	1,573	252	6:1

Thus, as we have seen, there are ten to twenty Negative Declarations issued for every EIS prepared.
Since 80-90 percent of all actions can be categorically exempted from EIS requirements, this means that
*less than 1 percent of all actions entail the prepar-
ation of an environmental impact statement.* Furthermore, evidence suggests that only a small fraction
of these actions (at *most* one-fifth of 1 percent) en-
counter delay, litigation, or other complications.[31]
Thus, even a vast majority of actions on which an
EIS is prepared receive routine approval following
the review process.

> 2. *The second major conclusion is that the
> costs of comprehensive environmental re-
> view, while often higher than they might
> be, are still generally insignificant
> when compared to other accepted planning,
> design, and regulatory costs (i.e., "in-
> vestment" costs).*

Although there is a great deal of variability
in overall state cost--from an estimated $30,000 per
year in Indiana to an estimated $30 million in Cali-
fornia--it has been the experience in the states
that, on the average, *the EIS process constitutes
only about .5 percent of project costs.*[32] At the
Federal level, this cost drops to as little as .1
percent of project costs reflecting the larger scale
of Federal projects.[33] A small-scale project may,
however, incur EIS-related costs in excess of 1 or
2 percent since there is a *minimum threshold cost* of
environmental review regardless of project size.
Thus, the small developer may be hardest hit by the
costs of environmental review, providing unfair
economies for large-scale development projects. This
fact is especially important in states such as Cali-

fornia and Washington where nearly 90 percent of all
environmental documentation takes place at the local
level.

Although the EIS process generally entails
costs of less than .5 percent of project cost, it is
clear that there are numerous factors which, histor-
ically, have resulted in unnecessary costs because
of excessive delay, litigation, or duplicated efforts. Some of these problems are inherent in the
EIS process. Many, however, can be mitigated by
certain adaptive changes; this is the topic of the
next section.

TOWARDS A MORE EFFICIENT AND EFFECTIVE EIS PROCESS

Many states have developed new and innovative
approaches for dealing with the problems of unnecessary delay, duplication, and overlap in the EIS process. Though relatively insignificant, the costs of
environmental review can be further reduced (or the
effectiveness of the process enhanced) by several
minor process changes which have the effect of
greatly streamlining certain aspects of EIS preparation and review. The following attempts to highlight
the innovative techniques being used in the eighteen
SEPA states and offers recommendations based on
these approaches.

Timing of EIS Preparation

A recent study by the U.S. General Accounting
Office (GAO)[34] showed that EIS's seldom cause long
delays in Federal public works projects. On the
other hand, EIS's frequently are not completed by
Federal agencies in time to be effectively utilized
in the decision-making process. Instead, they are
often completed late in project development, after
major decisions have been reached, and sometimes
after construction has begun.

As the GAO study noted:

The completion of EIS's late in project development can have serious consequences in two
respects. First, it tends to impede the document's basic purpose--surfacing environmental
impacts and alternatives to proposed Federal
actions for consideration along with economic
and technical factors when proposals are being
planned. In other words a late EIS tends to
relegate environmental considerations to lesser
importance than economic or technical ones,

51

contrary to NEPA's intent. Second, it risks
two types of project delays: (1) those associated with stopping projects to prepare late
EIS's, and (2) those due to replanning projects
or portions thereof, should the late EIS identify better alternative plans.[35]

There are two general reasons for late EIS preparation:

1. Many projects are "grandfathered," i.e.,
   even though location, design, or construction stages had been reached when the law
   became effective, it was determined that
   they still required EIS's; and
2. The procedures or practices of the initiating agencies permit late statement
   preparation.

Although a few of the states (most notably New York)
have attempted, through definition, to clarify the
"grandfathering" issue, it is the second problem
which can be dealt with most effectively.
     In order for an EIS to effectively accompany
a project proposal through an agency review process,
it must be completed during the stage of project decision making when the proposal is formulated. Project planning is the earliest stage of project decision making--the stage dealing generally with the
questions of (1) whether a project should be undertaken to meet a need or solve a problem, (2) where
the project should be located, and (3) what alternatives are available. The planning stage results
in a proposed action being recommended through an
agency's review process for approval. Other decision-making stages of design and construction follow project approval and deal generally with questions of how and when to implement a project.[36]
Thus, it is of critical importance that the implementing guidelines or rules and regulations for EIS
programs at any level of government deal effectively
with this issue. Figure 3.3 displays a prototype
project requiring an EIS and how the costs of environmental review are affected by the timing of impact statement preparation. By dovetailing the EIS
with early project planning, the process becomes
virtually indistinguishable from the baseline or
"normal" project cost curve. In fact, the EIS process may provide cost savings in certain cases where
problems are identified and avoided during the planning stage rather than during the construction stage

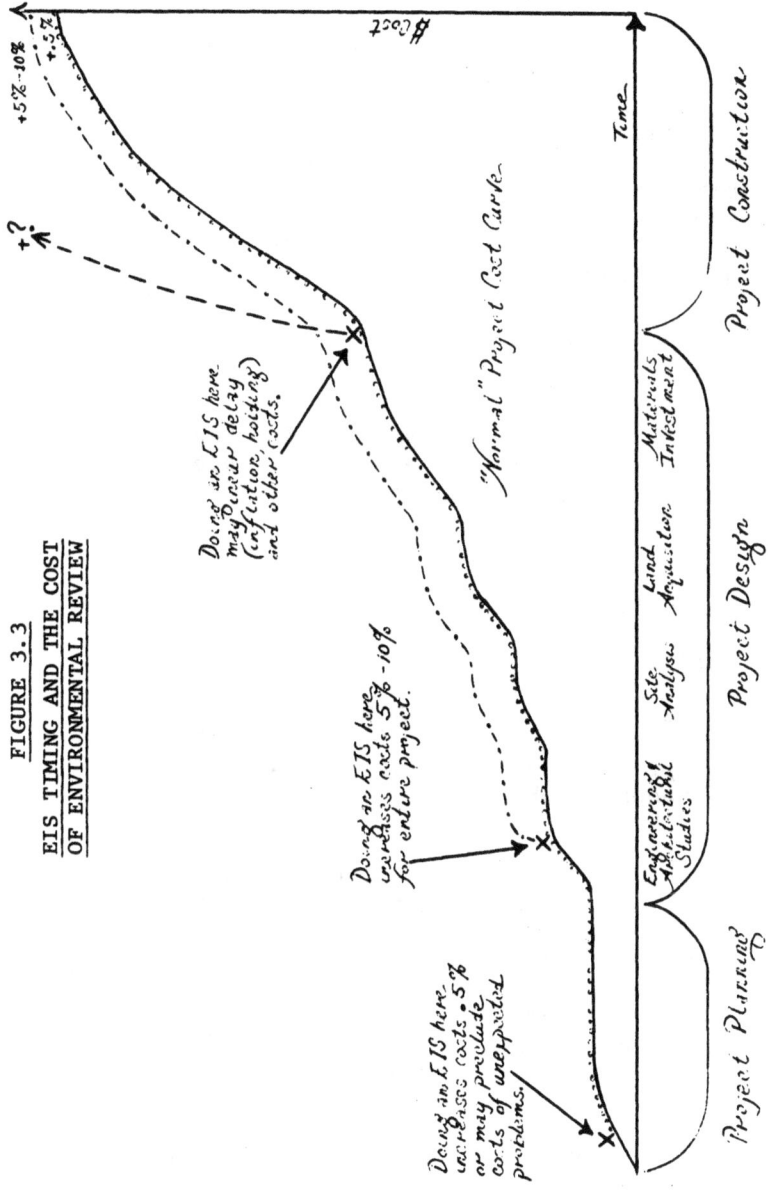

FIGURE 3.3
EIS TIMING AND THE COST
OF ENVIRONMENTAL REVIEW

Doing an EIS here increases costs .5% or may preclude costs of unexpected problems.

Doing an EIS here increases costs 5% - 10% for entire project.

Doing an EIS here may incur delay (inflation, holding) and other costs.

+?

+5% -10%
+.5%

$ Cost

"Normal Project Cost Curve"

Time

Engineering & Architectual Studies

Site Analysis

Land Acquisition

Materials Intrestment

Project Planning

Project Design

Project Construction

53

when they are astronomically expensive to correct.
The longer environmental review is delayed, however,
the greater the probability of costly duplications
of effort or delay due to litigation.

State programs have acted with varying degrees
of diligence concerning this matter. Although no
states impose any sanctions for negligence in this
regard, California and Washington dedicate a specific section of the rules and regulations to outlining this requirement while Hawaii, Indiana, Massachusetts, New York, Virginia, and Michigan incorporate equally strong statements into existing
formats. Maryland, Montana, and Utah address this
issue only tangentially and the remaining six SEPA
states appear, for one reason or another, to have
ignored the issue.

RECOMMENDATION #8

> *States should review their implementing guide-*
> *lines or rules and regulations to ensure they*
> *require that environmental impact statements*
> *be prepared as early as possible, preferably*
> *concurrent with project planning. Provision*
> *for early preparation of private EIS's (where*
> *applicable) should also be included.*

The section dealing with this issue in the recently
prepared draft regulations for NEPA implementation
is recommended as a model:

Section 1502.5 Timing
An environmental impact statement shall be
prepared as close as possible to the time an
agency makes a proposal (section 1508.21).
The statement shall be prepared early enough
so that it can practically serve as an important contribution to the decision-making
process and shall not be used to rationalize
or justify decisions already made (secs. 1500.2
(c), 1501.2, and 1502.2).
(a) For projects directly undertaken by
Federal agencies such statements shall be prepared at the feasibility analysis (go-no-go)
stage rather than the engineering design stage
(and may be supplemented at the latter stage
if necessary).
(b) For applications to the agency statements
shall be prepared at the latest immediately
after the application is received, but Federal
agencies are encouraged (preferably jointly

with applicable State or local agencies) to
prepare them earlier.

(c) For adjudication, the final environmental
impact statement shall precede the staff recommendation and public hearing.

Scope and Content of the Environmental Impact State-
ment

Heretofore, the only real guidance offered to
those responsible for EIS preparation has been in
the form of content requirements (either through
legislation or administrative regulation). These
provisions were described in the previous Chapter.
The unintended result of this open-ended process has
been the production of encyclopedic impact statements designed to withstand the test of legal challenge rather than short, concise impact statements
designed to deal with the important issues of specific projects.

In response to this lack of pre-EIS definition,
three states have adopted techniques to *limit the
scope and content of impact statements consistent
with the nature and magnitude of the proposed action
and the significance of its potential impacts.* These
techniques fall into two general categories: those
that categorically "focus" the content of all EIS's
and those that provide for the "scoping" of EIS content on a project-by-project basis.

Massachusetts has pioneered the technique of
early problem identification and policy conflict
resolution known as *scoping*. This process brings
those people likely to be concerned over a particular project together at the outset of the review
process in order to identify key issues. The following is excerpted from the Massachusetts statute:

> If an (environmental impact) report is required,
> the secretary with the cooperation of said per-
> son and agency shall, within the above-mention-
> ed thirty day period limit the scope of the re-
> port to those issues which by the nature and
> location of the project are likely to cause
> damage to the environment. The secretary shall
> determine the form, content, level of detail
> and alternatives required for the report. In
> the case of a permit application to an agency
> from a private person for a project for which
> financial assistance is not sought the scope of
> said report and alternatives considered therein
> shall be limited to that part of the project

55

which is within the subject matter jurisdiction
of the permit.

Thus, the Secretary of Environmental Affairs in Massachusetts has the power to determine the scope and
content of individual EIS's in consultation with interested and affected parties.

California has taken a more general approach to
the problem by legislatively *focusing* EIS scope and
content:

> 21102.1(2) The purpose of an environmental im-
> pact report is to identify the significant ef-
> fects of a project on the environment, to iden-
> tify alternatives to the project, and to indi-
> cate the manner in which such significant ef-
> fects can be mitigated or avoided. (Emphasis
> added)

In an attempt to further focus and streamline
the process, amendments to California's law states
that consideration of:

- the relationships between short-term uses
  of man's environment and the maintenance and
  enhancement of long-term productivity; and

- any significant irreversible environmental
  changes which would be involved in the proposed project should it be implemented--

shall be required only in EIR's prepared in connection with the following:

(a) The adoption, amendment, or enactment of a
    plan, policy, or ordinance of a public
    agency;
(b) The adoption by a local agency formation
    commission of a resolution making deter-
    minings;
(c) A project which will be subject to the re-
    quirement for preparing an environmental
    impact statement pursuant to the require-
    ments of the National Environmental Policy
    Act of 1969.

The aim of the amendments is thus the simplification
and streamlining of the EIR process by reducing both
the *number* of EIR's prepared (through early problem
identification and mitigation) and the *length* of the
report if it must be prepared (through focusing of
content).[37]

New York has embraced both the "scoping" and "focusing" concepts in recent revisions to its legislation and rules and regulations. Amendments to the law in 1977 added an EIS "focusing" clause similar to the one in California:

> Such statement (EIS) should be clearly written in a concise manner capable of being read and understood by the public, should deal with the specific significant environmental impacts which can be reasonably anticipated and should not contain more detail than is appropriate considering the nature and magnitude of the proposed action and the significance of its potential impacts.

Revised rules and regulations issued in January of 1978 provided for pre-EIS "scoping" sessions in a manner reminiscent of Massachusetts but with a stronger mandate for public participation:

> (h) Lead agencies shall make every practicable effort to involve applicants, other agencies and the public in the SEQR process. Consultants initiated by the lead agency can serve to narrow issues of significance and to identify areas of controversy, thereby focusing the subject matter of determinations and the scope and content of EIS's.

The President's Council on Environmental Quality has also adopted the "scoping" idea in its draft regulations for Federal environmental impact statements.

RECOMMENDATION #9

> *States should consider both "focusing" and "scoping" in their EIS programs as practical ways of eliminating extraneous information and concentrating on unresolved issues. Since the general "focusing" clause requires little to no additional manpower or expenditures, it is recommended to those states with limited access to resources for environmental review. For those states with existing in-house capabilities or access to such capabilities, it is recommended that the "focusing" clause be supplemented by a "scoping" process. Such a process may be optional (as in New York) but will probably be more effective in actually influencing*

EIS scope and content if made mandatory (as in Massachusetts).

In both cases, however, care must be taken not to circumvent the intent of comprehensive environmental review; issues must be carefully considered before being excluded from more in-depth analysis. It is also of vital importance that adequate provision be made for citizen participation in the scoping process. By involving interested, concerned, or affected members of the public *early* and in a *constructive* way, the chances of future litigation are greatly reduced. Thus, in the long run, "scoping" can save money as it greatly clarifies and shortens the entire environmental review process.

Coordination with Existing Requirements

1. The Permit Process

It is the purpose of the environmental impact statement process to assemble the often fragmented bits of information about proposed actions into comprehensive and coherent statements of overall impact. Since much of the data used in the preparation of EIS's is also necessary to comply with the various regulatory permit requirements, it is vitally important that the two procedures be coordinated.

Four states have chosen to deal with this problem by stipulating that *projects requiring permits or licenses are exempt from state EIS requirements* (see Table 3.5). This policy is stated straightforwardly in the laws and regulations of Indiana and Maryland. North Carolina and South Dakota, however, are somewhat less explicit on the issue. Although the South Dakota act states that EIS's are required for actions proposed or approved by state agencies, it also exempts several classes of action from the review process including "actions of an environmentally protective regulatory nature." Thus, permitted actions are exempted despite the fact that they constitute agency "approval." In North Carolina, projects requiring permits which do not involve public monies are exempt from the EIS requirement although localities are "authorized" to require impact statements for major private development projects.

Although the exemption of permitted actions from EIS requirements certainly removes any potential for overlap between the two functions, it seriously

TABLE 3.5
COORDINATION OF SEPA'S
WITH PERMIT PROCESSES

	"YES"		"NO"	
	Formally	Informally	Uncoordinated	Exempt
California	X			
Connecticut			X	
Hawaii				X^1
Indiana				X^2
Maryland				X^2
Massachusetts	X^3			
Michigan			X	
Minnesota	X^3			
Montana		X		
North Carolina				X^2
New Jersey		X		
New York	X^3			
South Dakota				X^2
Texas			X	
Utah		X		
Virginia	X^3			
Washington	X^3			
Wisconsin	X			

[1]Permits not issued on activities requiring EIS's.
[2]Projects requiring permits or licenses are exempt from EIS requirements.
[3]"Uniform Procedures" legislation either pending or in place.

jeopardizes the environmental review process since presumably, the largest projects (those subject to many regulatory requirements) escape truly systematic, comprehensive treatment. In an interesting turnabout, Hawaii has responded to this problem by *declaring those projects subject to EIS's exempt from the permit process.* Impact statement preparation and review is only informally (i.e., as a matter of consensus) coordinated with the permit process in Montana, New Jersey, and Utah, while in Connecticut, Michigan, and Texas, the two processes may actually be working at odds with one another since they are neither informally coordinated, nor formally exempted.

Eight states have attempted, in varying degrees, to *formally coordinate the two processes rather than to foster their mutual exclusion.* Through an "Environmental Permit Coordination Program" in Minnesota and an "Environmental Coordination Procedures Act" in Washington, completion of the EIS process is required prior to the issuance of individual state permits; in many cases, these permits may be approved on the basis of information provided in the EIS--thus, reducing duplication between the two processes. This technique also provides the applicant with a "master application process" through which all review and any required public hearings are coordinated through a *single* package application. New York is in the process of instituting such a coordination technique through a proposed "Uniform Procedures Act;" Massachusetts and Virginia currently apply the process to a limited extent--the former only on projects involving permits from the Massachusetts Department of Environmental Quality Engineering and the latter only on projects involving the issuance of permits among state agencies. In California, Massachusetts, New Jersey, and Wisconsin, the permit process has been generally coordinated with the EIS process through regulation but has not, as yet, been codified into a state "environmental coordination" type law. Wisconsin, however, has adopted the policy of excluding regulatory actions taken by the Department of Natural Resources from the EIS requirement.

An additional advantage to formally coordinating the impact statement process with permitting procedures is that the various time limits or "turn-around times" associated with each individual requirement can be combined into a single, more predictable set of time limitations for overall project review. Most states have suggested time periods within which the various steps of the EIS process should be accomplished. By establishing expeditious but reasonable time limitations on preliminary assessment and draft EIS preparation, *the EIS process can then run concurrently with other procedures relating to the review and approval of the action.* By allotting a certain period (e.g., thirty days) for public review, comment, and public hearings, agencies can then be required to act on all permit applications within a specific time after the filing of the final EIS. In this way, the EIS process may actually *expedite* rather than lengthen the overall time required for project review.

RECOMMENDATION #10

It is strongly suggested that states consider the formal coordination of the EIS requirement with the permit process through a comprehensive "Environmental Coordination" law. Such a process should contain provisions for allowing

> *the issuance of permits on the basis of rele-*
> *vant information provided in the EIS to reduce*
> *duplication and overlap between the two func-*
> *tions.*

2. Land-Use Regulation

An important consideration in implementing
EIS requirements in the states concerns how well
they relate to on-going governmental processes. *The*
role an EIS program plays depends, in large measure,
upon existing approaches to land-use planning and
regulation. This is especially true in those states
which require localities to implement environmental
review procedures.

Many believe that EIS programs might be re-
dundant in states with strong state-level land-use
regulation.[38] States such as Maine, Vermont, and
Oregon, for example, have thus far expressed little
interest in adopting state EIS requirements because
of their existing comprehensive land-use control
laws. Other states, however, such as Hawaii and
California, have taken strong steps in implementing
both land-use and discretionary environmental review
programs.

California, which has one of the nation's
strongest local planning laws, also applies the EIS
process to private projects subject to local government approval. As could be expected, the reaction to
this among California officials has been mixed:

> Some local planning directors in California
> express the opinion that they have had auth-
> ority for many years under the zoning and sub-
> division laws to require information about the
> environmental effects of proposed developments,
> and that the EIS program superimposes an un-
> necessary and burdensome requirement on the
> planning system. However, many planners feel
> that even with strong regulations and well-
> staffed local planning units, the EIS provides
> an additional and valuable means of forcing
> environmental factors to be taken into ac-
> count.[39]

New York now faces the monumental task of
implementing EIS procedures within the 62 cities, 62
counties, 931 townships, and 557 villages which have
been delegated the powers of land-use regulation.
Given the diversity in size and levels of sophistication among these localities, this is no small task.

State enabling legislation in the areas of zoning, subdivision regulation and site plan review, however, does not appear to present any major obstacles to such integration.[40]

Indeed, some states (including New York, Washington, California, and Minnesota) have extended this integration to its logical conclusion by *encouraging localities to prepare and submit environmentally sound comprehensive plans as "generic" or "programmatic" EIS's to the state for review.* Individual project proposals, in conformance with those plans, then need only provide specific supplemental information to comply with the EIS requirement. Used in this manner, the impact statement can become a vehicle for maintaining the integrity of existing comprehensive plans.

RECOMMENDATION #11

> *Since the environmental impact statement process represents an operational, case-by-case approach to many of the same goals intended by comprehensive planning, zoning, and other local land-use regulations, it is recommended that these two processes be used in combination as a system of environmental "checks and balances."*

To suggest that the comprehensive plan will envision and predetermine the kind and location of *all* potential activities seems naive; land-use policies will always leave a great deal of room for interpretation. Through the impact statement process, information about *specific* proposals can be gathered and analyzed in a systematic way and evaluated *against* the policies and standards of the local plans.

3. *The National Environmental Policy Act*

Traditionally, those states possessing their own mandates for environmental review (SEPA's) have eliminated duplication by waiving state requirements if a project must also comply with NEPA. Currently, only Connecticut, Indiana, New Jersey, Texas, and Utah exempt actions requiring EIS's under NEPA from state consideration. All other SEPA states accept information provided by the NEPA process *but require the Federal EIS to comply with state regulations* (i.e., be responsive to state and/or local concerns) before fulfilling SEPA requirements. Many encourage the preparation of *joint documents*--a process now

endorsed by the new CEQ regulations for NEPA.
Recently, the State of Massachusetts took this one
step further by directing the Secretary of Environmental Affairs to seek, from the Federal government,
the actual *delegation* of authority to carry out NEPA.

While the majority of SEPA states now require
Federal EIS's to be responsive to state needs, only
four states--Massachusetts, New York, North Carolina,
and Washington--explicitly address the issue of Negative Declarations prepared pursuant to NEPA. Since
preliminary assessments prepared by Federal agencies
understandably assess impacts from a national perspective, the determination of the potential for significant impacts at the state or local level can
only be made by a separate preliminary assessment
reflecting that perspective. As a result, the above
states do not accept Negative Declarations prepared
under NEPA. However, this problem may too be solved
by the new CEQ regulations for NEPA since Federal
agencies would be required to cooperate in fulfilling the requirements of state EIS laws as well as
discussing any inconsistencies with state or local
plans.

RECOMMENDATION #12

> *States should seek to coordinate the preparation
> of state and Federal environmental documents to
> the greatest extent possible, through joint
> planning, joint research, and joint preparation
> and review processes. Before any document is
> submitted in compliance with NEPA, however,
> states should require that it adequately address
> state and/or local concerns by also fulfilling
> the criteria of SEPA. This is especially im-
> portant as it relates to the preparation of pre-
> liminary assessments.*

Procedural and Substantive Streamlining

1. *Baseline Information*

A recent report by the Commission on Federal
Paperwork[41] examined the efficiency of using an "environmental resource inventory," (essentially a baseline data bank), for preparing environmental assessments on highway projects in Ohio. By using this
inventory to shorten the time and reduce the duplication involved in preparing the necessay preliminary
materials, it was found that about $480,000 in
highway construction funds (normally lost due to in-

flation) could be saved in the Dayton area alone. The inventory would cost only about $32,000 annually to maintain. Applied to the entire state, it was anticipated that the technique would yield the following benefits:

- The normal preparation time for environmental assessments could be reduced from between four to six and a half months to two to four weeks.

- The normal preparation time for all ODOT "programming" phase documents could be reduced from sixteen months to two months.

- Reducing the preparation and processing time by fourteen months could save about $6.5 million annually in inflation-incurred costs for Ohio's highway construction program.

- The paperwork burden on the local governments in Ohio would be reduced greatly.[42]

Although the potential benefits of such a system applied to an entire state environmental impact statement program seem staggeringly large, *only two of the eighteen SEPA states have made any real effort toward the systematic compilation of EIS-generated natural resource information.* While many states maintain files of environmental documents and review comments, only Wisconsin's Bureau of Environmental Impact maintains a library of EIS information while the state's Department of Natural Resources has developed a data processing system for storage and retrieval of environmental data. The state of Washington categorizes and summarizes all EIS's received or reviewed by the Department of Ecology into an "Environmental Impact Information System." Unlike Wisconsin, however, this system is primarily maintained as an information source within the agency. Although Hawaii has no data processing system currently in effect, it is the only state thus far to have included a provision in its EIS process allowing the incorporation of EIS-generated information into such documents by reference. New York, Massachusetts, Minnesota, Montana, and South Dakota all maintain EIS files but presently make no formal attempt to store the information systematically for quick retrieval and reuse.

64

> *States are encouraged to explore the possibili-*
> *ties of environmental resource inventories bas-*
> *ed on EIS-generated information as a way to*
> *avoid duplication in baseline data collection.*
> *States should also develop the means of in-*
> *corporating such information into EIS's by*
> *reference rather than repeating the inventory*
> *process for each document.*

Through the use of computers, data concerning geology, soils, topography, hydrology, flora, fauna, land-use, transportation, and utilities could be consolidated and coordinated into an easily useable system of storage and retrieval. In those states requiring EIS's only for state-sponsored actions, such a system could be established from the "top down" and centralized within a given state agency. Where EIS requirements extend to local governments, however, such information could be collected from the "bottom up" and stored at either the county or regional level.

2. *Exemptions*

Many actions (such as repair and maintenance of existing facilities, minor new construction, or other actions such as certain environmentally protective regulatory activities) can be safely assumed never to have significant adverse environmental effects, as long as they are not sited in areas of peculiar environmental sensitivity. Through the detailed listing of all such actions, it has been estimated that nearly 90 percent of all actions can be exempted from EIS requirements.[43]
Nevertheless, several states, including Connecticut, Maryland, Michigan, North Carolina, New Jersey, South Dakota, Utah, and Virginia have *neg-lected to develop detailed lists of exempt actions* (see Table 3.6). Virginia, however, specifically stipulates in its legislation that only projects costing in excess of $100,000 (excluding highway projects) are required to comply with the EIS requirement. Likewise, New Jersey's Executive Order exempts all projects costing less than $1 million unless they are proposed in environmentally sensitive areas. Connecticut's legislation exempts only "emergency measures" and "ministerial actions."
California, Indiana, Minnesota, and Washington all define, in their rules and regulations, more

specific classes of action which are to be considered exempt from the EIS process. Of these, Washington has developed the most intricate system of exemptions involving, in many cases, threshold levels
beyond which certain types of actions are subject to
environmental assessment. New York has taken a
slightly different approach by delineating both those
actions likely to require EIS's (Type I) as well as
those actions exempt from the requirements (Type
II). Massachusetts takes a somewhat firmer stand by
defining actions which are categorically *included* as
well as *excluded* from environmental assessment.

TABLE 3.6
EXEMPTIONS FROM THE EIS PROCESS

No Exemptions	Detailed Exemptions Classes of Actions	"Ministerial" Exemption
Connecticut*	California	California
Maryland	Hawaii	Connecticut
Michigan	Indiana	Massachusetts
New Jersey*	Massachusetts	Montana
North Carolina	Minnesota	New York
South Dakota	Montana	
Utah	New York	
Virginia*	Texas	
	Washington	
	Wisconsin	

*Legislation or Executive Order contains certain blanket
exemptions.

Probably of greatest utility, however, is the
system employed in Wisconsin and Texas. Actions are
classified as follows:

Type I: Actions which will *always* require
           EIS's.
Type II: Actions which *may* require EIS's
           given certain circumstances.
Type III: Actions which *never* require EIS's.

In this way, only "Type II" actions require preliminary assessments, thus eliminating this step for those
actions which will obviously require further environmental analysis.

A highly controversial question has been whether or not categorically to exempt all actions of a *ministerial* or non-discretionary nature. Indeed, many such actions (e.g., the issuance of dog licenses, marriage licenses, drivers' licenses, etc.) will obviously never entail significant environmental impact. However, it has also been contended that the issuance of building permits and similar local regulatory activities are ministerial since if all the requirements of the ordinances are met, there is no discretion on the part of an official to refuse the permit; therefore, the initiation of environmental review in such cases implies discretion which, it is claimed, does not exist.

While what constitutes a "ministerial" action and what constitutes a "discretionary" action is a matter of interpretation, eight states have faced this problem and gone in different directions. In Washington, Hawaii, and Minnesota, environmental considerations presumably must be taken into account in issuing building permits, grading permits and the like. This is especially critical in Washington where the EIS requirement is extended comprehensively to all local governments. The position of the Washington courts has been that where choice exists there is discretion and that local agencies may not waive the EIS requirement by claiming they are bound and limited by local laws:

> "(T)he fact that previous to SEPA the choice could be solely based on narrow or limited evaluative points set forth in an ordinance or statute is immaterial."44

An amendment to the Washington statute in 1973 reinforced these decisions by providing for a procedure to exempt certain classes of building permits for single family homes, adding that "building permits and acts not so classified shall not be presumed to either require or not require a 'detailed statement.'"

California, Connecticut, Massachusetts, Montana, and New York, however, have taken the opposite approach. As in the case of Washington, this is most critical in the states of California and New York where EIS requirements also encompass local government. By legislative amendment to the California act, EIS procedures do not apply to ministerial projects. Guidelines issued in 1973 interpreted "ministerial" to include:

1. Issuance of building permits;
2. Issuance of business licenses;
3. Approval of final subdivision maps;
4. Approval of individual utility service connections and disconnections.

Thus, it is possible that a small and ostensibly harmless industrial facility might require environmental analysis while a much larger and potentially more destructive subdivision could escape such review by virtue of its "ministerial" designation.

In New York, the statute excludes "official acts of a ministerial nature involving no exercise of discretion." New York regulations issued in January of 1978 simply state that a ministerial action is "an action performed upon a given state of facts in a prescribed manner imposed by law without the exercise of any judgment or discretion as to the propriety of the act, such as the grant of a driver's license." Since this definition is perhaps not very helpful, and the regulations encourage individual agencies to develop criteria for determining what categories of approvals are or are not ministerial, it will be interesting to see how the localities implement the ministerial exemption.

RECOMMENDATION #14

A detailed listing of all projects to be ex-
empted from the impact statement process, as
well as actions always requiring EIS's should
be developed by the Coordinating Agency. In
this way, the number of proposals requiring
preliminary environmental analysis can be great-
ly reduced. Agencies of state and, where ap-
plicable, local government should also be en-
couraged to develop their own system of exemp-
tions, subject to the review of the Coordinating
Agency. This should include a rider stipulat-
ing that general exemptions do not apply within
environmentally sensitive areas.

States should not categorically exempt "minis-
terial" actions from the EIS requirement both
because of the definitional problems and because
some actions defined as ministerial may involve
the potential for significant environmental ef-
fects. Individual ministerial actions (e.g.,
dog licenses, marriage licenses, drivers' li-
censes) which will never result in significant
environmental impact should be separately ex-
empted.

3. *Consultants and Fees*

Many states now provide for the contracting out of environmental work. New York, for example, states in its legislation that an EIS shall be prepared "by contract or otherwise" and in Massachusetts nearly all EIS's have been prepared by consultants. *Ultimately, however, all states require that the EIS be adopted by the sponsoring or lead agency as its own.* Thus, Washington, California, New Jersey, and Minnesota allow for the use of consultants, but all contracted work must be supervised by the lead agency.

It has been suggested that using consultants increases preparation costs due to consultants' higher staff salaries, and the multiplier used to cover staff fringe benefits, marketing expenses, company profit, etc.; the multiplier can run from 150 percent to 300 percent of staff salaries. Nevertheless, consultants provide expertise often not available in-house. Since EIS's often occur irregularly, consultants can also provide the numerous short-term staff needed. Unless an agency can forecast a continuous workload, adding the large numbers of full-time agency staff needed to do in-house work would most likely be inefficient. Consultant-prepared documents do tend to include more unnecessary information and be more elaborate than in-house documents. Vigorous supervision of consultants by the sponsoring agency, however, can do much to eliminate unnecessary costs.[45]

Since many of the states require EIS's on certain private projects entailing government approval, *the question of who should pay for document preparation and review--the state or the private applicant--also becomes a major concern.* Several states have instituted a "user-fee" for the purpose of recovering such costs. In California, for example, fees may be charged to recover the estimated cost of Negative Declarations and EIS's prepared for private applicants. The Montana law mandates that the state prepare EIS's for private projects and establishes a sliding scale charge based on total project cost; no fee is to be charged, however, until agency costs exceed $2,500. Amendments to Minnesota's law in 1976 established "chargeback rules." Projects costing under $1 million are exempt from this provision while the costs of EIS preparation for private projects in excess of $1 million can be charged back to the private proposer in accord with a formula based on project costs. The maximum re-

quired payment is .3 percent of the project cost
between $1-$10 million, .2 percent of the project
cost between $10-$50 million, and .1 percent of the
project cost over $50 million. New York has recent-
ly instituted a similar cost recovery provision; it
sets a ceiling of .5 percent of project costs and
stipulates that an applicant can be charged for document preparation or review but not both. A law
passed in 1975 required Wisconsin's Department of
Natural Resources (DNR) to charge a fee of .05 percent of estimated private project costs for EIS
preparation. However, because revenues derived from
this fee (anticipated at $1.5 million or more annual-
ly) would have gone to the general fund rather than
the agency, the law was altered in 1977 to allow
recovery of administrative costs only. The law was,
however, altered again in May of 1978 to allow DNR
to charge the full cost for EIS preparation on private projects.

RECOMMENDATION #15

> *States should institute a system of fees for
> recovering the costs of preparing and review-
> ing environmental documents for private appli-
> cants. Applicants should be assessed the
> actual cost of the work performed so long as
> the resultant fee does not exceed a reasonable
> percentage of project cost (such as .5 percent).
> The sliding scale system (i.e., charging a de-
> creasing percentage with increasing project
> cost) is not desirable because of its inadver-
> tent favoritism for large-scale projects. It
> is additionally recommended that public pro-
> jects include some provision for environmental
> review in their funding.*

4. *Statute of Limitation*

In the interest of predictability, most
states (as we have seen) are attempting to place
reasonable yet realistic time limits on the EIS process; this also has the effect of reducing the often
unnecessary but costly delays associated with environmental review. *Despite the increasing predictability of the EIS process itself, however, there
has not been an accompanying increase in the predictability of legal challenge to that process.* Indeed, only four states--California, Hawaii, Massachusetts, and Washington--have attempted to regularize the time within which citizen suit can be brought
against a decision rendered by the EIS process.

Perhaps most innovative, however, is the new limitations provision established in Massachusetts by amendments to the law in February of 1978. The most significant improvement over the original provision is the addition of a requirement to file a *notice of intention* to commence legal action pertaining to MEPA. Such a notice must be filed within sixty days of project clearance (or permit issuance in the case of private projects). This, it is hoped, will allow for a period of out-of-court problem solving before legal proceedings actually begin.[46] If problems are not resolved in this sixty day period, then litigation must commence within thirty days for a private project and within sixty days for a public activity.

RECOMMENDATION #16

> *States should adopt a comprehensive "Statute of Limitation" to limit and make predictable the timing and process of legal challenge to findings under the EIS laws.*

Postscript

The environmental impact statement process should not end with the approval of the proposed action. As the decision is implemented, the EIS can be supplemented with descriptions of mitigative measures actually taken; the agency should monitor the actual impacts of the decision after it is implemented and the EIS can be periodically updated to reflect those impacts. Such an environmental "post-audit" procedure has not yet been attempted, but it has great potential. By comparing actual impacts with predicted impacts, agencies can greatly refine their impact-prediction techniques. Monitoring of actual impacts would also provide invaluable information which could guide agencies in revising and refining their guidelines, standards, and procedures. The extended EIS thus becomes a feedback mechanism to guide future decisions.[47]

The EIS should be viewed as a complete record of an agency's decision-making procedures: from the inception of the proposal through the search for alternatives and mitigative measures, the projection of unavoidable adverse environmental impacts, the presentation and balancing of competing environmental and non-environmental values, input from other agencies and the public, the justification for the final decision, the record of subsequent mitigative

71

measures, and the actual impacts of the action as implemented. Although it is entirely unreasonable to expect that the faithful application of the EIS process will always result in the "best" or "correct" decision, it should, at the very least, provide us with a logical progression of well-documented steps through which we can learn to make better decisions.

Part II
The Information

4

Focus: Environmental Review in Six States

This Chapter contains detailed information about the environmental impact statement programs in six states--California, Washington, North Carolina, Massachusetts, Minnesota, and New York. In-depth, personal staff investigation in these states resulted in a level of information and analysis sufficient to warrant separation from that of the other twelve SEPA states (Chapter 5) for which no on-site investigations were conducted.

The information on each state is divided into three main sections:

1. A background statement covering program structure and evolution;
2. An in-depth treatment of EIS program cost and caseload. This section begins with a statewide orientation and closes with specific information about the EIS Coordinating Agency and selected Project-Originating Agencies; and
3. A summary of recent adaptations (amendments or rules changes) designed to improve the effectiveness or efficiency of the EIS program.

Detailed citations for legislation or regulation cited in the text have been omitted as these may be found in the appropriate SEPA "Fact Sheet" in Appendix A.

CALIFORNIA

California was the first state to establish an EIS requirement patterned after NEPA. The California Environmental Quality Act (CEQA), which was signed into law on September 18, 1970, recognized the need to understand the "relationship between the maintenance of high quality ecological systems and the general welfare of the people of the state."

As enacted in 1970, CEQA consisted of a declaration of legislative findings and policy and two operative chapters requiring environmental impact reports to be prepared on state and local *projects* "which could have a significant effect on the *environment* of the state." (NEPA, it should be recalled, required EIS's on "legislation and other *major* Federal *actions* significantly affecting the quality of the *human* environment." Also, the term "environmental impact report" is used in the California act; the words "environmental impact statement" are usually used to describe the document required under the Federal law.)[48]

Although the content of environmental impact reports (EIR) as described in CEQA's Chapter 3 was modeled after that of NEPA, the California Act went one step further:

> 21100. All state agencies, boards, and commissions shall include in any report on any project they propose to carry out which could have a significant effect on the environment of the state, a detailed statement by the responsible state official setting forth the following:
>
> (a) The environmental impact of the proposed action.
> (b) Any adverse environmental effects which cannot be avoided if the proposal is implemented.
> (c) Mitigation measures proposed to minimize the impact.

(d) Alternatives to the proposed action.

(e) The relationship between local short-term uses of man's environment and the maintenance and enhancement of long-term productivity.

(f) Any irreversible environmental changes which would be involved in the proposed action should it be implemented.

(Emphasis added)

The addition of the "mitigation entry," although not an entirely new concept, did represent a needed clarification in legislative thinking. An amendment to CEQA in 1972 added a seventh entry to the above list:

(g) The growth-inducing impact of the proposed project.

The significance of this addition lies in its implicit mandate for consideration of *indirect* as well as direct impacts.

CEQA mandated the responsibility of "coordinating" the state impact statement requirement to the Office of Planning and Research (OPR). Section 21103 as originally enacted stated that OPR was responsible for coordinating the establishment of objectives, criteria, and procedures for the preparation and evaluation of EIR's. The fact that the law, as originally enacted failed to provide clear authority and a deadline for issuance of detailed guidelines, however, caused CEQA to be virtually ignored for the first two years.[49]

The Friends of Mammoth vs. Board of Supervisors of Mono County[50] case, however, precipitated a series of changes in the Act--both in substance and responsibility. In that decision, handed down in September of 1972, *the California Supreme Court held that the law applied not only to governmental actions but also to the approval by government of private activities which could have a significant effect on the environment.* It had been assumed by most observers that private projects were not included within the EIS requirement but the Court noted that NEPA was used as a pattern for the California Act and that Federal guidelines would require an EIS under similar circumstances. Therefore, the Court concluded that under the California law, state and local agencies must file an EIS before acting on private projects.[51]

The Friends of Mammoth decision resulted in confusion over the status of private projects already in progress, and this caused considerable alarm in the construction industry, the building trades and the financial community. Numerous lawsuits were threatened (nearly 250 lawsuits were actually filed pertaining to CEQA between 1970 and 1975), and some officials called for amending the law so that it would not apply to private projects.[52] The following describes the situation:

> Intensive negotiations among legislators, developers, and environmentalists led to the enactment of a compromise measure in December of 1972. This measure added a number of clarifications and new provisions to the 1970 law, but it had little effect on the most important features of the act, including the Court's interpretation. It also granted retroactive exemption to projects already built or underway, and imposed a 120-day moratorium on implementation of the EIS requirement.[53]

The 1972 amendments required the Office of Planning and Research to develop, and the Secretary of Resources to adopt, the statewide guidelines (equivalent to rules and regulations) for the administration of the Act by early 1973. This effectively delegated the substantive responsibilities of coordination to the Resources Agency and left the Office of Planning and Research as primarily a Clearinghouse and research unit.[54]

The guidelines were developed and subsequently amended through an open hearing process and state agencies and localities were required to adopt their own rules consistent with those of the state. Although many aspects of implementation were clarified by the 1973 guidelines, the approach taken to the *exemption issue* was probably of fundamental importance. In addition to the categorical exemption of actions such as repair or replacement of existing facilities, the guidelines also provided a blanket exemption for all actions considered to be "ministerial" in nature:

15073. MINISTERIAL PROJECTS.
(a) Ministerial projects are exempt from the requirements of CEQA, and no environmental documents are required. The determination of what is "ministerial" can most appropriately be made by the particular public agency involved based upon its analysis of its own laws, and each public agency should make such determination either as a part of its implementing

regulations or on a case-by-case basis. (P.R.C. 21080(b)).

(b) In the absence of any discretionary provision contained in the relevant local ordinance, it shall be presumed that the following actions are ministerial:

    (1) Issuance of building permits.
    (2) Issuance of business licenses.
    (3) Approval of final subdivision maps.
    (4) Approval of individual utility service connections and disconnections.

(c) Each public agency should, in its implementing regulations or ordinances, provide an identification or itemization of its projects and actions which are deemed ministerial under the applicable laws and ordinances.

(d) Where a project involves an approval that contains elements of both a ministerial action and a discretionary action, the project will be deemed to be discretionary and will be subject to the requirements of CEQA.

This provision, although clear in its intent to streamline the process of environmental documentation, holds the potential for ambiguous and inconsistent determinations of EIS threshold. For example, it is possible that a small and ostensibly harmless industrial facility could be "captured" under the Act while a much larger and potentially more destructive subdivision could escape such review by virtue of its "ministerial" designation.[55]

The following documents are required by the California Environmental Quality Act (CEQA) and its administrative guidelines for the purposes of assessing environmental impacts of proposed projects and providing a record of those assessments. These documents combine to form a *screening process* through which only those projects of potentially adverse impact must undergo a full evaluation. Projects of insignificant environmental consequences can thus be certified as such and either follow a less extensive process of evaluation or be excused entirely from further analysis:

1. Notice of Exemption - A statement of record, for categorically exempt, ministerial, and emergency projects, all of which are exempted from environmental analysis.
2. Initial Study - An identification of potential adverse impacts for all projects not exempted from environmental analysis.

3. Negative Declaration - A statement of rec-
    ord, for all non-exempt projects found to
    be without potential or real significant
    adverse environmental impact.
4. Environmental Impact Report (EIR) - An eval-
    uation of projects which may have a signifi-
    cant adverse impact.
5. Notice of Completion - A statement of record
    for all projects for which EIR's have been
    drafted, notifying interested and respon-
    sible parties that a draft EIR has been com-
    pleted and is available for review and com-
    ment.
6'. Notice of Determination - A statement of
    record for all projects with negative dec-
    larations or EIR's notifying all interested
    and responsible parties of the project's
    environmental impact and the final action
    taken on it.
7. Statement of Overriding Consideration - A
    statement of record for approved projects,
    found to result in substantial adverse im-
    pact, indicating overriding social and/or
    economic bases for their approval.

Despite this formidable system of documentation,
there is no enforcement authority in the state administration that can require compliance with the act.
There are no administrative sanctions for failure to
file an EIR, nor for failure to implement the CEQA
process. The regulations and procedures adopted by
state agencies and local governments under CEQA are
not reviewed administratively; thus, enforcement is
made the responsibility of the political process and
of individuals or groups who may challenge an agency
decision in court for non-compliance with CEQA.

COST AND CASELOAD

Statewide Implementation

A number of attempts have been made to assess
the overall costs as well as general effectiveness of
the EIS process in California. Most notable, however,
has been a series of reports prepared for the Assembly
Committee on Local Government entitled The California
Environmental Quality Act, An Evaluation.[56] Through
analysis of the California EIR Monitor (a bi-weekly
announcement of environmental documentation) and the
records of the State Clearinghouse, estimates of CEQA-
related activity and its associated cost were develop-

ed at the state as well as local level.

The CEQA Evaluation Reports analyzed costs in terms of four principal components:

1. the cost of document preparation, review, and administration;
2. delay costs;
3. costs associated with uncertainty in project execution created by CEQA; and
4. mitigation costs.

In addition, costs were separated to the extent possible in terms of (1) public costs on public projects; (2) public costs on private projects; and (3) private costs on private projects (see Table 4.1).

It must be pointed out, however, that the accuracy of the figures derived from this study are questionable for several reasons, including:

● The calculation of local government CEQA costs were made on the basis of a "representative" sample of twenty cities and counties throughout the State. Extrapolation from samples is, of course, always questionable.

● State agency costs were calculated, by and large, on the basis of estimated incremental costs attributable to CEQA in preparing environmental documents.

● The costs of delay were again based on sample averages in reaching discretionary decisions through the CEQA process.

● The costs of mitigation and uncertainty were estimated from a sample of 185 EIR's.

Given the complexities of delay, mitigation, and uncertainty costs, however, it is unlikely that a more precise estimate of CEQA-related costs would be of great utility in judging the cost-effectiveness of the Act.

Approximately 4,000 EIR's were prepared by government agencies in 1974. Of these, 3,400 were prepared by cities and counties (see Table 4.2). Thus, nearly 90 percent of CEQA environmental documentation takes place at the local level. There is also evidence that agencies are issuing increasing numbers of Negative Declarations as a result of being able to work with applicants during the design of a project, thereby obtaining changes which effectively mitigate potential adverse environmental impacts.[57]

TABLE 4.1
THE COSTS OF CEQA
in millions of dollars
Estimated for FY 1975-76

| | COSTS TO: | | | |
GOVERNMENTAL LEVEL	PUBLIC ON PUBLIC PROJECT	PUBLIC ON PRIVATE PROJECT	PRIVATE ON PRIVATE PROJECT	TOTAL
Cities, counties and special dist.	$ 9.6	$ 7.2	$10.0	$26.8
State (doc. prep.)	4.0	--	--	4.0
State and certain special dists. for review and administration	1.0	2.0	--	3.0
	$14.6	$ 9.2	$10.0	$33.8

TABLE 4.2
APPROXIMATE NUMBER OF EIR'S
PREPARED ANNUALLY BY TYPE
OF JURISDICTION

Cities	2,200	58%
Counties	1,200	32%
Special Districts	200	5%
State Agencies	200	5%
TOTAL	3,800	100%

The document preparation costs for all state agencies attributable to CEQA was estimated at $4 million per year. Appendix B-1 presents a breakdown comparing document costs to actual project costs for each agency.

The cost of environmental review of other agency documents by state agencies and certain special regional agencies was estimated at $1 million. A breakdown for this cost component may be found as Appendix B-2. In addition, there are administrative expenses associated with the Clearinghouse, the Attorney General's Office, and agency coordination. These costs were estimated at $2 million. This total of $3 million was allocated between public and private projects on the basis that approximately 67 percent of all environmental documents are associated with private projects.[58]

Public costs on private projects
$3,000,000 x .67 = $2 million

Public costs on public projects
$3,000,000 x .33 = $1 million

Thus, a total yearly state cost of $7 million was seen as attributable to CEQA as compared to $16.8 million for local government and $10 million for private applicants.

When allowances for project delay, uncertainty, and mitigation were included in this estimate, however, the costs changed dramatically (see Table 4.3). With these cost components included, the in-

TABLE 4.3
ESTIMATED APPROXIMATE INCREMENTAL
COSTS OF IMPLEMENTING CEQA
in millions of dollars
FY 1975-76

CATEGORY OF COST	FOR PRIVATE PROJECTS		FOR PUBLIC PROJECTS
	COST TO PRIVATE APPLICANT OR CONSUMER	COST TO PUBLIC	COST TO PUBLIC
Document prep., review, and administration	$10 mil	$9 mil	$15 mil
DELAY — Inflation Effect	$30 mil impact on price	--	Less than $11 mil impact on price
DELAY — Carrying or foregone opportunity	$21 mil	None	Less than $11 mil
Uncertainty	Perhaps $2 mil due to proj. abandonment remainder unknown	None	Unknown but believed to be low
Mitigation	Less than $2.5 mil	Unknown but believed low	Unknown but may be substantial

cluded, the incremental costs of implementing CEQA were estimated to be in the range of *$50 million to $75 million per year for the state as a whole.* This represented approximately 0.5 percent of project costs and the total costs of environmental analysis attributable to CEQA were estimated at between $2-$3 per capita annually.[59]

With this general picture in mind, we turn to an analysis of selected state agencies in an attempt to get an idea of the costs involved in specific roles related to CEQA.

The Resources Agency

The California Resources Agency includes the various departments, boards, and commissions in state government that have responsibilities for natural resources management and regulation (see Appendix B-2). The agency as a whole qualifies as a major originator of projects (one such department--Water Resources--will be focused on in the next section).

Amendments to CEQA in 1972 required that the Secretary of Resources issue statewide guidelines for the administration of the Act. Although no special "Office of Environmental Impact Review" has been established to coordinate CEQA review, a "coordination unit" from the Department of Water Resources has been expanded to encompass both CEQA as well as NEPA review. The unit now collects and consolidates the review comments of the various state agencies on EIR's prepared under CEQA as well as EIS's under NEPA. The following cost breakdown (see Table 4.4), therefore, reflects these coordination responsibilities under CEQA. It must be stressed, however, that the Office of the Secretary does not incur any of the review costs associated with the various departments within the Agency aside from those of the "Coordination Unit."

Department of Water Resources

DWR, a department within the Resources Agency, is responsible for the study and coordination of water development projects undertaken by counties, cities, state agencies, public districts, and the Federal Government. It also initiates projects of its own and since 1972, has completed five EIR's and between forty and fifty Negative Declarations.[60] Most of the projects requiring EIR's have been massive in scale and will be dealt with in more detail below.

82

```
                        TABLE 4.4
                     RESOURCES AGENCY
                  OFFICE OF THE SECRETARY¹
                   Estimated FY 1977 Cost

Assistant Secretary (2/3 at $2,500/mo.)          20,000/yr.
NEPA-CEQA Coordinator (1/2 at $2,400/mo.)        14,400/yr.
1.5 Secretarial Positions at $1,024/mo.          18,000/yr.
²Training Program, Handbooks, Public Info
 Coordination Unit (Staff)                       60,000/yr.
  3 clericals at $1,024/mo.                       36,000/yr.
³Legal Counsel and Litigation                    negligible
 EIR Monitor                                     12,000/yr.
 Hearing Costs                                    5,000/yr.
 Indirect Costs - 15% salary                     23,000/yr.
                                                $178,400/yr.
```

[1]All figures were established through time and salary estimates of expended cost. Salary and other figures were extracted from the Governor's Budget FY 1977. The breakdown does not include direct costs.

[2]This function is filled by the Office of Planning and Research (OPR). Thus, not all activities associated with *coordination* are borne by the Resources Agency. Although OPR is now primarily a Clearinghouse, it does occasionally, in addition to its public information function, get involved in substantive review.

[3]The Secretary's Office participates in very little legal action. Most litigation associated with CEQA is in other areas such as Forest Practices of Water Resources. In fact, the Attorney General's Office maintains a *9-person staff* at salary costs of nearly *$260,000 per year* specifically to handle environmental matters including CEQA and NEPA.

The fact that DWR utilizes a *program budget* approach greatly facilitates the costing out of CEQA activities. For example, the Department's overall budget (including capital outlay) for 1976-77 was approximately $291 million. A sequential breakdown of those costs related to CEQA is displayed in Table 4.5.

DWR anticipates that the workload under CEQA will continue to increase thus making additional staff necessary to review EIR's and prepare department, agency and State comments.

TABLE 4.5
DEPARTMENT OF WATER RESOURCES[1]
Estimated Cost

	1975-76	1976-77	1977-78
1. Preparation and processing of environmental documents	–	$ 32,100	$ 33,000
2. Provide advisory and training assistance to all Department units in the preparation and processing of environmental documents.	–	32,100	33,000
3. Continue environmental education program to advise and aid Department employees, other public employees, and anyone else the Department comes in contact with on understanding how various Department and agency activities relate to the environmental assessment process.	–	16,000	16,400
4. Update and expand Handbook on Environmental Planning Guidelines.	–	16,000	16,500
SUBTOTAL[2]	$ 91,349	$ 96,300	$ 98,900
5. Review of other agencies' EIR's[3]	439,339	547,200	575,900
6. EIR preparation[4]	⌐300,000	⌐300,000	⌐300,000
TOTAL	$830,988	$943,500	$974,800

[1]All budget figures contain a State overhead allowance which includes direct and indirect costs, legal fees, (incurred largely by Agency attorneys) and litigation costs.

[2]Program activities under this component consist primarily of procedures development, review and staff advice.

[3]Of this figure, approximately $96,000 per year is charged to the Secretary's Office for funding the previously mentioned "Coordination Unit." Also, in the case of Federal projects, some of the cost is attributable to NEPA compliance.

[4]The cost of EIR preparation is separately calculated below. It is anticipated that the incremental cost of CEQA will eventually approach zero as it becomes increasingly interwoven into the planning process. For the present, however, the cost is as follows:

    (a) The Peripheral Canal Project - Over $1 million has been spent already on environmental documentation for this controversial development which extends back into the 1960s. Expenditures per year are estimated at $200,000.

84

(b)   An average of *three* other on-going projects requiring EIR's at an approximate cost of *$20,000-$30,000*
per year for each.

By summing components (a) and (b), we derive the average
annual figure of $300,000 used in the table.

CalTrans

CalTrans is California's equivalent of a State
Department of Transportation. The name was adopted
primarily as a means of distinguishing the agency
from its Federal counterpart.[61]   Unfortunately,
agency responsibility regarding NEPA and CEQA is somewhat ambiguous.   Since most highway projects involve
Federal funding (and thus compliance with NEPA), the
incremental cost attributable to CEQA is difficult
to specify.

CalTrans has integrated the Federal and state
EIS laws by utilizing Federal Highway Administration
(FHWA) guidelines to satisfy NEPA and then adding
the "growth inducing impacts" and "mitigative measures" requirements to achieve CEQA compliance. Thus,
the preparation of a single document practically
satisfies both laws.

Last year, CalTrans had an operating budget of
approximately $400 million with an additional $400
million allotted to construction. The cost of EIS/
EIR preparation is paid for by the state in normally
all cases except Interstate projects (the state could
pursue Federal monies for environmental documentation
on *all* Federal aid projects if it so desired).   Appendices B-1 and B-2 document the costs and preparation times of selected EIR's prepared by CalTrans.

Although the following cost breakdown makes no
attempt to distinguish between NEPA and CEQA compliance, it is useful to the extent that it portrays
the proportions which environmental documentation can
reach relative to overall agency costs:[62]

FY 1975-76 - 490 person-years = $12.5 million
FY 1976-77 -~350 person-years = $9-10 million

The above figures represent the *overall* cost of NEPA/
CEQA compliance and include administration and management, document preparation, review of incoming EIS's,
legal counsel and litigation, and all areas of direct and indirect expense.

Most environmental documentation work takes place
within a designated "environmental unit." Of the 245
positions which have been authorized for this unit
statewide, only 230 are currently filled. Thus, the

additional person-years shown above reflect time
spent in review by *people in other units*. The drop
in manpower and cost from 1975-76 to 1976-77 reflects
changing program directions (from an emphasis on *construction* to one of *renovation*).

*Thus, the cost of complying with NEPA and CEQA
ranges somewhere between .5 and 1 percent of project
cost and 1.5 to 2 percent of the total yearly budget.*

COST EFFECTIVE ADAPTATIONS

Since the initial amendments to CEQA in 1972,
the law has been qualified seven more times. Probably of the greatest significance with respect to
cost effectiveness, however, are the most recent
amendments of September 1976; they reflect the way
in which perceptions and hence program priorities are
changing as experience with CEQA increases. The
first such amended passage represents an attempt to
define the functional role of "mitigation measures"
in environmental review--a departure from the idea
of using such measures to "balance" the adverse effects:

> 21002. The Legislature finds and declares that
> it is the policy of the state that public agen-
> cies should not approve projects as proposed if
> there are feasible alternatives or feasible
> mitigation measures available which would sub-
> stantially lessen the significant environmental
> effects of such projects, and that the pro-
> cedures required by this division are intended
> to assist public agencies in systematically
> identifying both the significant effects of
> proposed projects and the feasible alternatives
> of feasible mitigation measures which will avoid
> or substantially lessen such significant ef-
> fects. The Legislature further finds and de-
> clares that in the event specific economic,
> social, or other conditions make infeasible
> such project alternatives or such mitigation
> measures, individual projects may be approved
> in spite of one or more significant effects
> thereof.

In order to achieve the objectives set forth in
the above, the stated *purpose* and *scope* of the environmental impact report (EIR) itself is modified:

> 21102.1(2) The purpose of an environmental im-
> pact report is to identify the <u>significant</u>

effects of a project on the environment, to
identify the alternatives to the project, and
to indicate the manner in which such significant
effects can be mitigated or avoided.
(Emphasis added)

In an attempt to focus and streamline the process
further, the amendments state that the consideration
of:

- the relationship between short-term uses of
  man's environment and the maintenance and enhancement of long-term productivity; and
- any significant irreversible environmental
  changes which would be involved in the proposed project should it be implemented

shall be required only in EIR's prepared in connection
with the following:

(a) The adoption, amendment, or enactment of a
    plan, policy, or ordinance of a public
    agency.
(b) The adoption by a local agency formation
    commission of a resolution making determina-
    tion.
(c) A project which will be subject to the re-
    quirement for preparing an environmental
    impact statement pursuant to the requirements
    of the National Environmental Policy Act of
    1969.

The aim of the amendments is thus the simplification
and streamlining of the EIR process by reducing both
the *number* of EIR's prepared (through early problem
identification and mitigation) and the *length* of the
report if it must be prepared (through focusing of
content).[63]

WASHINGTON

The State Environmental Policy Act (SEPA) of
1971 followed the precedent set by NEPA very closely
in legislative intent as well as procedural style.
The policy section of the Act is virtually identical
to the "Environmental Bill of Rights" espoused in
NEPA.

The impact statement prerequisite established by
the law applies to all "state agencies, municipal and
public corporations, and counties" and only "major
actions significantly affecting the quality of the
environment" are required to comply with the full ar-
ray of environmental documents. This, too, is largely reminiscent of NEPA as are the areas of consideration included within the impact statement:

   (i)   the environmental impact of the proposed
        action;

  (ii)   any adverse environmental effects which
        cannot be avoided should the proposal be
        implemented;

 (iii)   alternatives to the proposed action;

  (iv)   the relationship between local short-term
        uses of man's environment and the main-
        tenance and enhancement of long-term
        productivity; and

   (v)   any irreversible and irretrievable com-
        mitments of resources which would be in-
        volved in the proposed action should it
        be implemented.

Unlike NEPA, however, the Washington Act specified no administering agency comparable to the President's Council on Environmental Quality (CEQ). Thus,
the courts were forced to play a major role in clarifying the intent of the legislation. Overall the
State Courts have acted vigorously to enforce the
statute in cases of private as well as public development. Like California's EQA, SEPA has also survived
(and ultimately been strengthened) by an amending
process which followed the Friends of Mammoth decision. *Amendments to the law in 1974 made it clear*

that permits were potentially subject to the EIS requirement.

Under the 1974 amendments, a "Council on Environmental Policy" (CEP) was created with rule-making power to develop definitive statewide EIS guidelines. Localities were required to adopt their own rules consistent with the state guidelines. As directed by the Legislature, the Council dissolved on June 30, 1976 and its powers, duties, and functions were transferred to the Department of Ecology (DOE) where an Environmental Review Section was set up to coordinate the SEPA program as well as certain agency activities under NEPA. The Department was also directed, by the Legislature, to promulgate a model ordinance.

Although amendments to Washington's Environmental Policy Act have certainly been significant, the interpretation of the Act through statewide guideline preparation has provided the most tangible statement of policy direction. The guidelines adopted by the Council on Environmental Policy in December of 1975 included many significant additions to and clarifications of the Act. *Probably paramount in importance, however, was their development of categorical exemptions.* The CEP delineated such exemptions very *specifically*--an approach which is fundamentally different from the more broad designation of general classes of action used in a number of other SEPA states (e.g., California, New York). Thus, for example, under the general class of "Enforcement and Inspection," the guidelines exempt "any action undertaken by an agency to abate a nuisance or to abate, remove or otherwise cure any hazard to public health or safety" but specify that "no license shall be considered exempt by virtue of this subsection; nor shall the adoption of any ordinance, regulation or resolution be considered exempt by virtue of this subsection."

The Council intentionally avoided the categorical exemption of such arbitrary designations as "ministerial actions." This was done in recognition of the fact that actions of a so-called "non-discretionary" nature may indeed have a significant effect on the environment."[64] As one Washington Court put it:

> The change in the substantive law brought
> about by SEPA introduces an element of dis-
> cretion into the making of decisions that were
> formerly ministerial, such that even if we
> assume arguendo, that the issuance of a grad-
> ing permit was, prior to SEPA, a ministerial,
> nondiscretionary act, SEPA makes it legislative
> and discretionary.[65]

Other important clarifications contained in the original guidelines included a special allowance for *environmentally sensitive areas* designated at the local level and the addition of a requirement to discuss the "mitigative measures" proposed to minimize impacts in EIS's. Finally, the guidelines mandated the establishment of "SEPA Public Information Centers" by all state agencies for the purpose of maintaining and making available copies of all Negative Declarations and EIS's filed by the agency. Local agencies were also encouraged to establish regional SEPA Public Information Centers on a county-by-county basis.

After a private application has been received or a program proposal defined, the first step in the EIS process is to designate a lead agency. Disputes involving lead agency determinations are to be resolved by the Department of Ecology within fifteen days. The review process then runs as follows:

Lead Agency Responsible Official Makes Threshold Determination (Less than 15 Days)

- Evaluate checklist initially without request for further information.
- If necessary, more information may be required of applicant or consultation request may be sent to other agencies with jurisdiction.
- Will there be significant adverse impacts?
- If non-significant adverse impact: Prepare and list a final Declaration of Non-Significance (Neg. Dec.). Copy to SEPA Information Center.
  Except: Prepare and circulate (15 days) a "proposed" Neg. Dec. where other agencies and demolition or clearing/grading permits are involved. List with SEPA Information Center.
  For a "proposed" Neg. Dec. List, wait 15 days and either: finalize, opt for EIR, or gather more information. (If an agency with jurisdiction formally challenges a "proposed" Neg. Dec., it must prepare an EIS.)
- If significant adverse impact: Prepare and list a Declaration of Significance (copy to SEPA Information Center). An EIS will be needed.
  NOTE: Project planning and modification to mitigate or avoid impacts can reverse a "significant impact" decision to "non" and a Neg. Dec.

If Significant Impact, Lead Agency Prepares an EIS.
- If necessary, request "consultation" from other

- agencies with jurisdiction or expertise.
- Simultaneously release <u>draft EIS</u> and list with SEPA Information Center.
- A SEPA <u>public hearing</u> must be held if: Lead Agency decides or 50 people request or 2 jurisdictional agencies request.
  1. Any SEPA hearing will be held not less than 15 days nor more than 51 days after draft EIS release/listing.
  2. The hearing notice will include AP & UPI only if action is regional or statewide. Otherwise, only area newspaper.
  3. The hearing notice will occur no less than 5 days prior to hearing.
- Prepare and list <u>Final EIS</u> (within 75 days of listing/releasing draft).
  1. If no critical comments on draft: circulate that result as the final.
  2. If critical comments received on draft, either:
     a. Prepare a document with comments and responses, <u>or</u>
     b. Rewrite the draft with comments, responses, disagreements.
  3. Where possible, use or supplement a prior SEPA statement.
  4. Where possible, substitute or supplement a final EIS under NEPA.
  5. Distribute Final EIS and list with SEPA Information Center.

Although the Environmental Review Section of DOE is responsible for coordinating SEPA activities, it is not vested with the power of rejection vis a vis impact statement review. Final review authority resides in the Courts and, to a lesser extent, in the granting of *permits* which is accomplished through the Environmental Coordination Procedures Act. Indeed, the condition or denial of permits constitutes a given agency's "power of rejection" over private activities. With respect to public projects, however, in which the originating agency prepares its own impact statement and makes the final review decision, the public disclosure aspects of SEPA are the only real lever for assuring public satisfaction with agency decisions.[66]

COST AND CASELOAD

Statewide Implementation

The 1974 amendments to SEPA required the Office
of Program Planning and Fiscal Management (OPPFM) to
submit a report to the 1977 legislative session on
the implementation status of the Act.[67] The requirements were stated as follows:

> Each state agency, political subdivision,
> municipal and public corporation, and county
> shall review all actions taken to implement
> this chapter (the State Environmental Policy
> Act) and may submit a report of such actions
> to the Office of Program Planning and Fiscal
> Management, which shall compile and analyze
> such data and prepare a report which shall be
> submitted to the forty-fifth regular session
> of the legislature. In addition, information
> on the cost of implementation and administra-
> tion of the act shall be included in such re-
> port including the cost of preparation of all
> detailed statements since the effective date
> of this 1974 amendatory act.

Needless to say, the cost of overall SEPA implementation proved a very elusive figure and although the
OPPFM study attempted such a calculation, the statistical basis for its estimates was less than conclusive:

GOVERNMENTAL UNITS CONTACTED	REPLIES RECEIVED	PERCENT RETURN
39 Counties	18	46
267 Cities	84	31
42 State Agencies	29	69
22 Community Colleges	4	18
55 Port Districts	24	44
30 Public Utility Districts	11	37
455 Units	169	37

Numerous governmental units chose not to respond to the questionnaire designed jointly by OPPFM
and DOE. Given the permissiveness of the legislative requirement, however, the data appeared to go
far in providing the order-of-magnitude information
desired. The utility of the estimates is further
bolstered by the fact that although only 37 percent
of the total units reported, only one large unit
failed to respond (King County).

Table 4.6 contains the summary cost data tabulated from the OPPFM analysis.

TABLE 4.6
THE COSTS OF SEPA[1]
Costs by Type of Reporting Unit

Total, all reporting cities	$2,486,805
Total, all reporting counties	330,952
Subtotal, all reporting cities & counties	$2,817,757
Total, all reporting state agencies including four-year institutions of higher education	$5,971,721
Total, all reporting community colleges	650
Total, all reporting port districts	246,037
Total, all reporting public utility districts	956,805
Grand Total	$9,992,970

Costs by Type of Activity[2]

Total, Guidelines Development Activities	$ 418,098
Total, Impact Assessment Activities	2,301,901
Total, Statement Preparation Activities	7,272,971
Grand Total	$9,992,970

[1]The data above was derived from a wide range of cost estimating techniques (from accounting records to educated guesses, and from incremental costing to actual costing).

[2]The costs shown under Guidelines Development are essentially one-time costs while those shown under Impact Assessment and Statement Preparation are continuing.

The highest cost reporting units were the Department of Transportation ($4.9 million), City of Seattle ($1.8 million), Washington Public Power Supply System ($.7 million), and the Department of Natural Resources ($.4 million). Appendices B-3 through B-5 contain the breakdown of state agency costs as compiled by OPPFM. It must be remembered, however, that these figures encompass the period of May 1974 through October 1976. Thus, any approximation of SEPA costs for a one-year period would have to be derived as a proportion of this 28-month period:

$$\frac{\$10 \text{ million}}{28 \text{ months}} = \frac{\text{Annual Cost (x)}}{12 \text{ months}}$$

Estimated Cost of
SEPA   = $4,300,000 per year
(x)

93

It is significant that while 80 percent of SEPA documentation takes place at the local level (since 1972, local agencies have prepared 1,072 draft EIS's compared to 100 by state agencies), less than half of the cost is incurred within the cities and counties (see Table 4.7). Local costs, however, are by no means negligible.

TABLE 4.7
DRAFT EIS PREPARATION
UNDER SEPA

Year	Local Agencies	State Agencies	DOE*	Federal Agencies (NEPA)
1972	139	11	6	25
1973	260	32	6	18
1974	317	22	8	45
1975	188	26	11	36
1976	168	19	12	30
5-Year Total	1,072	110	43	154
1977 (thru Sept.)	182	9	6	32

*Included with in the "State Agency" figures.

A recent study by the Washington State Association of Counties[68] estimated that it costs approximately $250,000 simply to put people in place in the counties for SEPA implementation. In order to operationalize the system, this figure would have to double which means that county-level implementation costs at least $½ million per year. This does not include the considerable expense incurred by localities in legal fees and procedural advice.

A similar study is being conducted by the Washington Association of Cities. Although more accurate figures are forthcoming, the estimated cost of SEPA implementation within cities and towns is at least that of the county cost. *Thus, the lower range for total local SEPA costs (excluding legal counsel and indirect costs) is on the order of $1 million per year.*

With the benefit of this statewide cost overview, we now turn to analysis of the costs of specific state agency functions under SEPA.

94

Department of Ecology

As discussed previously, the guidelines established by the Council on Environmental Policy remand-
ed the responsibilities of SEPA coordination to DOE
upon its dissolution in 1976. Aside from its duties
of guidelines revision and EIS review, however, the
agency must also prepare impact statements by virtue
of its regulatory function and thus takes on certain
characteristics of a "project originating" agency.
(In fact, six of the nine draft EIS's prepared for
SEPA by state agencies as of September of 1977 were
prepared by DOE.)

In order to fulfill these multi-faceted responsibilities, an Environmental Review section was created within DOE and 6.5 persons were added, spending
between one-half and three-quarters of their time on
SEPA-related matters.[69] Disciplines of the new staff
included environmental science, community affairs,
and civil engineering. Success in fully implementing
the duties under the Act and the CEP guidelines, however, has been somewhat hampered because of insufficient staff and budget levels.

The following cost breakdown (Table 4.8) is derived for a one-year period using the estimates provided by DOE for the Office of Program Planning and
Fiscal Management's assessment of SEPA. Appendix
B-6 contains the actual data provided by DOE for the
period of May 1974 through October 1976.

It is estimated that the cost of preparation
of the average EIS is $5,000. Applying this
to the six impact statements prepared in the first
eight months of 1977 yields a figure of about $30,000
(see Table 4.9). This appears to align quite closely with the $41,000 projected for EIS preparation
over the course of a full year. By subtracting the
above figure from the total, we obtain a *SEPA coordinating cost of approximately $110,000 per year.*

Department of Transportation

Washington State DOT's relationship regarding
SEPA is somewhat different from most other state agencies. Since most highway projects involve Federal
funds, and thus require a NEPA statement, the incremental cost attributable to SEPA is difficult to dis-
cern. Furthermore, given the magnitude and controversy of most highway projects, very few Negative
Declarations are issued and the length of the resultant EIS often mirrors the scale of the project. Although there were only six draft EIS's begun in 1977

TABLE 4.8
DEPARTMENT OF ECOLOGY*
Estimated FY 1976 Cost

[1]Admin./Mgmt.	$ 7,500
[2]EIS Review	12,500
[3]EIS Prep	41,000
Local Assistance	25,000
Guideline Questions	16,640
Petition Responses	12,500
SEPA Inf Register	4,160
[4]Legal Counsel and Litigation	1,500
(.05 x 30,000)	
Applicants Brochure	2,000
Guideline Preparation	2,000
Model Ordinance	3,000
Indirect Costs (20% salary)	22,360
	$150,100

*The breakdown does not include direct costs nor does it include the cost of time spent by various department people (air, water resources, etc.) in supplying information, reviewing EIS's for the Environmental Review Section.

[1]Administration and management constitute about 10-15 percent of the stated salary costs.
[2]This includes the review of an estimated ninety NEPA statements. Additionally, DOE received only 70-75 percent of all draft SEPA EIS's developed in the state.
[3]This cost is not part of DOE's "coordination costs." It reflects the "project originating" aspects of the Department by virtue of its regulatory function.
[4]An estimate of 5 percent of one lawyer's time at $30,000 per year reflects the small amount of litigation associated with DOE.

(and only thirteen in the last three years), the estimated cost of in-house EIS preparation staff alone is in excess of $500,000 for the year.[70]

Although the following statistical breakdown (Table 4.9) includes the joint cost of NEPA and SEPA compliance, it is significant to the extent that it suggests the order of magnitude that such costs can reach. All figures were established through the use of time and salary estimates and other estimates of expended costs.

The figure of $1.5 million per year for joint SEPA-NEPA compliance represents a lower limit. In fact, the OPPFM report lists the cost at nearly $5 million for the 28-month period of its study (see Appendix B-3). OPPFM also requested a tally of the

```
                     TABLE 4.9
             DEPARTMENT OF TRANSPORTATION
               Estimated FY 1976-77 Cost

Admin./Mgmt.                          $   111,200
EIS Prep. Staff                           541,000
EIS Review Staff                          213,400
Consultants                                90,000
Clerical Staff                             56,000
Hearings                                   45,500
¹Legal Counsel and Litigation              60,000
Printing and Dist.                         18,000
Travel                                      2,000
Office Space                               23,000
Equipment, Materials, Telephone             5,000
Indirect Costs (20% salary)               185,000
                                      $1,450,500  ~1.5 million

     ¹This figure represents the salaries of two full-time at-
  torneys from the Attorney General's Office assigned to
  DOE for SEPA-related conflicts.  The cost of litigation
  is impossible to discern but is estimated to be at least
  $150,000 in addition to the above.
```

costs of those things called for in SEPA that are
not required in NEPA. Although DOT was unable to
provide such information at the time, it has since
been able to identify the following separable costs:

1. SEPA Public Information Center
    SEPA guidelines require that each agency
    establish a SEPA public information center
    and make available a series of project
    status registers and files. The cost of
    this requirement to the Department is ap-
    proximately $18,000 per year.
2. SEPA Environmental Checklist
    SEPA requires that a specific environment-
    al checklist be prepared for many minor
    projects which are exempt from the NEPA
    process that have been determined to have
    insignificant impact on the environment.
    The additional cost for this requirement
    is estimated to be $36,000 per year.
3. The Department is required to review num-
    erous minor project environmental check-
    lists which are prepared by other agencies.
    The estimated cost of this activity is
    $7,000 per year.

4. <u>Local Agencies' SEPA Environmental Documents</u>
The Department is also required to review approximately 500 project environmental documents per year which are prepared by local agencies. The estimated cost of this activity is $176,000 per year.

TOTAL $237,000 per year

It is significant that the proposed guidelines provide the ability to defer to an adequate Federal impact statement but do not provide a similar deferment for preliminary assessments performed under NEPA. Although the additional cost of $36,000 for SEPA checklist preparation is small in comparison to DOT's overall cost of compliance, it constitutes the only major duplication of effort between SEPA and NEPA.

COST EFFECTIVE ADAPTATIONS

Amendments to the Act in 1974 established a *statute of limitation* concerning environmental review and SEPA compliance. This was an important step towards the institution of regulatory consistency:

RCW 43.21C.080 Notice of any action taken by a government agency may be publicized by the acting government agency, the applicant for, or the proponent of such action...Any action to set aside, enjoin, review, or otherwise challenge any such governmental action for which notice is given...on grounds of noncompliance with the provisions of this chapter (the SEPA Act) shall be commenced within <u>thirty</u> days from the date of last newspaper publication...(and) any subsequent <u>governmental</u> action on the proposal for which notice has been given...shall not be set aside, enjoined, reviewed, or otherwise challenged on grounds of noncompliance with the provisions of RCW 43.21C. 030(2)(a) through (h) (the impact statement provision) unless there has been a substantial change in the proposal...or unless the action now being considered was identified in an earlier detailed statement or declaration of nonsignificance as being one which would require further environmental evaluation. (Emphasis added)

98

Between 1975, when the original SEPA guidelines were adopted, and mid-1977, petitions were received by both the Council and the Department of Ecology suggesting changes of various sorts. As a result, guidelines revision was undertaken by DOE to simplify language, smooth procedures, and reduce unnecessary paperwork. Following a series of public hearings, revised guidelines were adopted December 31, 1977. Some of the more important revisions include the following:

1. An exemption for all renewal activities which do not contain major alterations in design.
2. An emphasis on short analytical impact statements. This goal is encouraged, in large measure, by the Environmental Impact Information System within DOE. The system provides for the categorization and summarization of all EIS's received or reviewed by the agency and is used as an information base for future EIS preparation. Also proposed was the ability to defer to adequate Federal Negative Declarations but this was later reversed.
3. The definition of "Action" (for clarity's sake) as something which may adversely modify the physical environment only. (This assumes that a project having the potential for social or economic impacts will be large in scale and thus have certain physical impacts associated with it as well.)
4. The continued refinement of threshold levels and exemptions. For example, in the 1975 guidelines, an exemption for weed control along right-of-ways was granted as follows:

(f) Periodic use of chemical or mechanical means to maintain a utility or highway right-of-way in its design condition: PROVIDED, That chemicals used are approved by the Washington State Department of Agriculture and applied by licensed personnel.

The proposed guidelines, however, take this one step further:

(f) Periodic use of chemical or mechanical means to maintain a utility or highway right-of-way in its design condition (:-PRO-VIDED,-That) except when this occurs in

99

watersheds which are controlled for the
purpose of drinking water quality, so long
as the chemicals used are approved by the
Washington State Department of Agriculture
and applied by licensed personnel.

5. The deletion of the regional SEPA Public
   Information Centers and increased efforts
   upon upgrading the responsiveness of a
   centralized information center within DOE.

NORTH CAROLINA

The 1971 enactment of the North Carolina Environmental Policy Act (NC-EPA), made North Carolina the sixth state to pass environmental impact assessment legislation. An interesting aspect of NC-EPA was its time limitation--*the Act was originally enacted as a two-year experiment slated to expire on September 1, 1973.* Because a key legislative committee decided that the Act needed a more extensive test period, the Legislature voted in 1973 to extend the law until 1977, rather than make it permanent;[71] the 1977 session again voted to reinstate the Act until August 1, 1981.[72]

In the mold of NEPA, NC-EPA requires that state agencies comply with the goals of the state's environmental policy and prepare statements assessing the impact of their actions which might significantly affect the environment:

> Section 4(2) Any state agency shall include in every recommendation or report on proposals for legislation and actions involving expenditure of public moneys for projects and programs significantly affecting the quality of the environment of this State, a detailed statement by the responsible official setting forth the following:
> a. The environmental impact of the proposed action;
> b. Any significant adverse environmental effects which cannot be avoided should the proposal be implemented;
> c. Mitigation measures proposed to minimize the impact;
> d. Alternatives to the proposed action;
> e. The relationship between the short-term uses of the environment involved in the proposed action and the maintenance and enhancement of long-term productivity; and
> f. Any irreversible and irretrievable environmental changes which would be involved in the proposed action should it be implemented.

101

In the absence of clear legal directive, *it has been unofficial state policy that the EIS requirement was not intended to cover the issuance of private permits where no public money is involved.*[73] NC-EPA does, however, allow for the inclusion of private actions in the EIS process by *authorizing* (but not mandating) localities to require EIS's for major development projects:

> §113A-8. Major development projects. The governing bodies of all cities, counties, and towns acting individually, or collectively, are hereby authorized to require any special-purpose unit of government and private developer of a major development project to submit detailed statements, as defined in G.S. 113A-4(2), of the impact of such projects.

The term "major development projects" includes but is not limited to "shopping centers, subdivisions and other housing developments, and industrial and commercial projects but shall not include any projects of less than two contiguous acres." This authorization has, however, gone largely unheeded--only one county and one municipality have passed such ordinances. Indeed, the state has failed to encourage local implementation; no suggested guidelines for localities have as yet been issued.[74]

The North Carolina Council on State Goals and Policy was initially charged with reviewing impact statements and evaluating state environmental goals. Following a change in Administration in 1972, however, responsibility for overall coordination of NC-EPA was delegated to the Department of Administration (DOA). Until early 1976, the only guidelines for implementing NC-EPA issued by DOE were in the form of a Memorandum dated February 18, 1972. The memo directed each state agency to establish its own regulations for implementing the requirements of the Act in conformance with the general criteria contained in the memorandum. This directive was essentially ignored by all agencies except the Department of Natural and Economic Resources (DNER) which promulgated its own guidelines in 1976. Furthermore, only DNER and the State Department of transportation have prepared impact statements under NC-EPA and only DNER has filed the actual series of reports required for compliance under §113-6 of NC-EPA.[75]

New NC-EPA Guidelines[76] were promulgated by the Department of Administration effective February 1, 1976. Under these guidelines, the State Clearinghouse, located within the Department of Administration's Office of Intergovernmental Relations, coordinates the review of both state and Federal EIS's. Prior to the adoption of the new NC-EPA guidelines, there was no explicit requirement that Negative Declarations be prepared where no EIS was deemed necessary. Consequently, most agencies generally made such determinations internally. Public notice of the availability of draft and final EIS's has, however, been published along with a brief project description, in the bi-weekly North Carolina Environmental Bulletin.

Although the decision of whether or not to prepare an EIS is left entirely to the discretion of the initiating agency, the first step in the new EIS procedure established in the guidelines is a "determination of significance." Such a determination is made on the basis of an "environmental assessment questionnaire" which is circulated through the State Clearinghouse for agency comment. The guidelines establish the following general criteria as indicators of project significance:

> .0-05 ENVIRONMENTAL ASSESSMENT QUESTIONNAIRE
> To facilitate the determination of "significance" an Environmental Assessment Questionnaire was developed. Reviewing agencies will use this questionnaire as a means to determine "significance" if any of the following conditions exist:
> (1) A proposed project or program has the potential to degrade the quality of the environment, to curtail the range of the environment or to achieve short-term to the disadvantage of long-term environmental goals;
> (2) The possible effects of a project are individually limited but cumulatively considerable especially when such effects are growth-inducing; or
> (3) The environmental effects of a project will cause substantial adverse effects on human beings either directly or indirectly.

This procedure is to be followed, however, only in those instances where "an official of a project or program questions the need for an EIS." Otherwise, the first opportunity for Clearinghouse and public review of an agency action follows submission of the actual draft EIS or Negative Declaration. In an effort to ensure that such submission occurs before agencies are committed to any particular course

of action, the guidelines stipulate that: "prior to
the release of any funds other than design funds, an
assessment of the potential environmental impact of
the proposed action must be completed."

Copies of comments received on the draft EIS's
or Negative Declarations are forwarded by the State
Clearinghouse to the project sponsor, upon completion of a thirty day review period. The project initiating agency is then supposed to "review the environmental effects of the action in light of the
comments received" and address these comments in the
final statement. Copies of the final EIS, with comments attached, are then required to be sent by the
Clearinghouse "to all interested parties and agencies
from whom comments were solicited.

*Conspicuously absent from the North Carolina
Environmental Policy Act regulations is any mention
of public hearings or meetings.* Thus, the extent to
which public hearings or meetings are held either in
plan preparation or on the draft EIS itself depends
entirely on independent agency initiative. While
the regulations do allow for appeal by private individuals or public agencies of an environmental
finding or statement, the hearing provided for on
such an appeal is not an open one. If a hearing is
determined to be necessary, a hearing officer is required to be appointed by the Governor or his designee "to review the objections and present recommendations for resolution to the Governor for action."
The applicable section from the legislation is as
follows:

> 113A-5. Review of agency actions involving
> major adverse changes or conflicts. Whenever,
> in the judgment of the responsible State of-
> ficial, the information obtained in preparing
> the statement indicates that a major adverse
> change in the environment, or conflicts con-
> cerning alternative uses of available natural
> resources, would result from a specific pro-
> gram, project or action, and that an appropri-
> ate alternative cannot be developed, such in-
> formation shall be presented to the Governor
> for review and final decision by him or by such
> agency as he may designate, in the exercise of
> the powers of the Governor.

*Thus, under NC-EPA, the authority for policy
decisions involving projects with major environmental adversities rests with the Governor or his designee;* a differentiation is made between project

advocate and project evaluator which has the potential for more impartial judgment of environmental effects. Until the administrative and regulatory agencies of North Carolina take affirmative steps to implement the Act, however, the potentials of NC-EPA will remain unrealized.

COST AND CASELOAD

Statewide Implementation

The use of the North Carolina Environmental Policy Act as an integrated planning tool has simply not occurred. The Act was essentially ignored from the time of its enactment until the issuance of guidelines in 1976. Since then, the EIS requirement has been invoked only on a sporadic basis; the following figures attest to this:

TABLE 4.10 ENVIRONMENTAL DOCUMENTS PREPARED PURSUANT TO NC-EPA		
	FY 1975	FY 1978
EIS's	7	2
Environmental Assessments	8	22

Since the bulk of this activity has come from two agencies of state government, we turn to these agencies for a closer examination of the costs incurred in their implementation of NC-EPA.

Department of Natural and Economic Resources

The Department of Natural and Economic Resources (DNER) has been one of only two agencies (Department of Transportation being the other) predisposed to implementing or utilizing the environmental protection activities established by NC-EPA. DNER has both developed departmental regulations for EIS's and filed EIS's while most other state agencies have done neither. This comparatively high compliance with the letter and spirit of NC-EPA has made DNER, rather than DOA, the "lead agency" with respect to the implementation of the Act.

Recognizing the expansion of its mandate as a result of NC-EPA, DNER added an Environmental Assessment element in the Fall of 1971. The "Environmental Assessment Unit" of DNER is involved primarily

in the *review* of environmental impact statements and
special studies. As of May 1976 DNER has directly
prepared only two environmental impact statements in
order to comply specifically with NC-EPA. On the
other hand, in just the first five months of 1976,
the number of documents reviewed by DNER reached
eight environmental impact statements/assessments
and seven hundred A-95 notifications in application
for Federal assistance. The A-95 notifications are,
of course, reviewed under the mandate of or response
to the National Environmental Policy Act (NEPA)
rather than NC-EPA. Most of the EIS's reviewed by
DNER are also Federal or NEPA-related documents. *All
state agencies combined prepared only seven formal
state (NC-EPA) EIS's in 1975.*

Excluding two "Environmental Assessment Coordinators," the staff time for review of environmental
documents is borrowed from either DNER divisions or
seven regional offices. The designation of "environmental contact person" in the divisions and regions
is official, but its function is in addition to or
subordinate to each staff member's formal, salaried
departmental role. The "contact people" essentially
assume the added responsibility of environmental review without receiving any financial compensation or
lessening of other responsibilities.

The following cost figures (Table 4.11) were
calculated using time estimates from individual
state agency reviews and information from the Environmental Assessment Unit with DNER.[77] Time estimates could only be given for fiscal 1976 as previously, EIS's and Negative Declarations were sent
to anyone within the departmental divisions for a
statement of concurrence. The cost of EIS Administration (guidelines, working papers and organizing
conferences) is generated only by two "Environmental
Assessment Coordinators."

It should be stressed that the Environmental
Assessment Unit does not show up as a line item on
the Departmental Budget; it is rather, a "hidden item"
under the Office of the Secretary. As a "hidden item,"
the Environmental Assessment Branch makes no budget
requests to the Department and the Department makes
no line item or program requests for environmental
assessment to the Governor and the State Legislature
Environmental Assessment subsequently receives no
direct legislative appropriation; whatever operating
funds the Branch utilizes come from allocations to
the Department's Administrative Budget. To be exact,
the funds come from the Administrative Services Budget and amount to no more than enough to cover the

```
                        TABLE 4.11
          DEPARTMENT OF NATURAL AND ECONOMIC RESOURCES
                     Estimated FY 1976 Cost

Staff                                  Cost of EIS review FY76

12 "Contact People"                          $11,798
Environmental Assessment Coordinator (I)
  - Administration                               458
  - EIS review activity                        1,070
Environmental Assessment Coordinator (II)
  - Administrator                                458
  - EIS review activity                        1,070
Environmental Assessment Secretary               391

Total Cost                                   $15,240
```

This figure reflects the cost of staff time for:
- review of all environmental assessments under the state Act.
- review of all state EIS's.
- preparation of any assessments and EIS's (3 EIS's have just been submitted by DNER but they do not show up in the caseload tally).
- review of some Federal EIS's as it was difficult for the personnel contacted to separate out their time. Also, most of the state's EIS's that are circulated for review are joint Federal/State transportation projects.

The above figure does not reflect the cost of:
- travel.
- soft and hard supplies for state EIS office activity.
- the staff time for attending an occasional environmental assessment committee meeting.
- contractor time (one EIS done in FY75 by DNER was contracted out and cost about $35,000).

salaries of the two Environmental Assessment Coordinators (approximately $35,000 annually) in the Environmental Assessment Branch.

The Administrative Services Budget has covered the two salaries for several years and expects to cover them in the near future. However, the Administrative Program of DNER has proposed some changes for the 1977-79 Biennium Plan. The Department is requesting the State Legislature to allocate $34,873 for fiscal 1977-79 and $33,797 for fiscal 1978-79 for the addition of one community development planner, one clerk-steno, and various supplies and materials.

The special request is obviously an attempt to increase staff capability; the request is also considered to be a test of the Legislature's and the state's attitudes towards environmental impact assessment *since it marks the first time DNER has asked directly for funds for EIS activities.* The outcome will have strong implications for whether or not environmental assessment activities will continue to remain "hidden items" within DNER.

North Carolina Department of Transportation

DOT is the only other state agency to have implemented NC-EPA. Since all state Departments of Transportation must comply with NEPA (where Federal money is involved) the costs attributable to state-level EIS requirements are difficult to separate. The North Carolina DOT, however, offers a unique opportunity since so little environmental documentation has been prepared pursuant to NC-EPA. Cost figures were calculated using time estimates given in person-hours for each individual on the staff of the Environmental Unit within DOT. The figures estimate the preparation costs of one state EIS and one state Negative Declaration (see Table 4.12).

TABLE 4.12
DEPARTMENT OF TRANSPORTATION
Estimated Document Costs

	1 N.D.	1 EIS
Supervisor		
Highway Engineer 3	$ 168	$ 336
Project Engineer		
Highway Engineer 2	1,539	3,078
Highway Engineer 1	838	1,676
Technician 2 (drafting)	15	33
Highway Engineer 3	30	67
Wildlife Biologist 3	120	239
Community Value Specialist	223	446
Highway Engineer 2 - Air	120	246
Highway Engineer 2 - Noise	120	246
Noise Technician 3	95	194
Noise Technician 2	65	134
Clerical Staff	26	52
Borrowed Staff total	224	448
TOTAL	$3,583	$7,128

The salaries used were in the middle of the
base salary range for that position. This was done
so that estimates would apply for both fiscal years
as the level of experience fluctuated.[78]
In fiscal 1975, fifteen Negative Declarations
were prepared; most of them were never filed with
the State Clearinghouse, however, since the Division
of Highways believed it need not go through a formal
review process. Three EIS's were done; two went on
file in the State Clearinghouse. The total estimate
of staff time for fiscal 1975 was $75,129.
In fiscal 1976, seven Negative Declarations were
prepared and one EIS was completed under the State
Environmental Policy Act. The one EIS that was done
was extremely brief--not more than 100 person-hours
were spent on it. The estimate of total staff time
for fiscal 1976 was $25,687.
These estimates are believed to be an underestimation for these reasons:

1. The current salaries of the Environmental
    Unit total $155,120.
2. The caseload for fiscal 1975 represented
    58.5 percent of the total Negative Declar-
    ations and EIS's prepared under both En-
    vironmental Policy Acts. This attributes
    a cost of $90,745 to NC-EPA.
3. The caseload for fiscal 1976 represented
    26.3 percent of the total Negative Declar-
    ations and EIS's prepared under both En-
    vironmental Policy Acts. This attributes
    a cost of $40,900 to NC-EPA.
4. Some of the staff time commitment is les-
    sened by use of consultants.
5. The cost estimates do not include adminis-
    trative overhead, hearings, travel, soft
    and hard supplies, printing and circulation,
    and review of the few incoming EIS's, all
    of which are costs applicable to the state
    EIS program within the Environmental Unit
    of the Department of Transportation.

Thus, the estimates should be increased by 10-16 percent to account for the costs not included. Probable range of costs attributable to the Department
of Transportation for the state EII program is,
therefore, *$82,000-$86,500 for fiscal 1975 and
$28,000-$29,500 for fiscal 1976.*

COST EFFECTIVE ADAPTATIONS

Since there has been little effort to implement the Act, the costs involved have been miniscule. As a result, there has been little impetus for creative interpretation of the Act beyond the initial issuance of Guidelines in 1976.

MASSACHUSETTS

The Massachusetts Environmental Policy Act was signed into law on July 18, 1972. Section 61, which went into effect on December 31, 1972, required "all agencies of the Commonwealth" to review and evaluate all of their activities, to determine the impact on the natural environment of such activities, and to use "all practicable means and measures to minimize damage to the environment." The key phrase, "damage to the environment," was defined as:

Any destruction, damage or impairment, actual or probable, to any of the natural resources of the commonwealth and shall include but not be limited to air pollution, water pollution, improper sewage disposal, pesticide pollution, excessive noise, improper operation of dumping grounds, impairment and eutrophication of rivers, streams, flood plains, lakes, ponds, or other surface or subsurface water resources; destruction of seashores, dunes, marine resources,underwater archaeological resources, wetlands, open spaces, natural areas, parks, or historic districts or sites. Damage to the environment shall not be construed to include any insignificant damage to or impairment of such resources.

While Section 61 mandated the review, evaluation, and mitigation of environmental impacts associated with agency activities, Section 62 (which became effective on July 1, 1973) established the vehicle for accomplishing these goals--the environmental impact report (EIR) process. The legislation defined the character of an EIR in terms similar to that of NEPA:

An environmental impact report shall contain detailed statements describing the nature and extent of the proposed work and its environmental impact; all measures being utilized to minimize environmental damage, any adverse

111

short-term and long-term environmental consequences which cannot be avoided should the
work be performed; and alternatives to the
proposed action and their environmental consequences. The preparation of said report
shall be commenced during the initial planning
and design phase of any work, project, or
activity subject to this section and the report shall be so prepared and disseminated as
to inform the originating agency, reviewing
agencies, the appropriate regional planning
commission, the attorney general and the public
of the environmental consequences of state
actions and the alternatives thereto prior to
any commitment of state funds and prior to the
commencement of the work, project, or activity.

Court decisions involving NEPA established an
early precedent for the extension of EIS requirements to permitted actions as well as
those activities directly undertaken or financed by Federal agencies--a precedent later
formalized in the CEQ guidelines. It was on
the basis of this precedent (and the experience
in California) that MEPA was informally defined as covering:[79]

1. projects or activities of state agencies,
    projects or activities financed with state
    monies, or projects and activities of lo-
    cal redevelopment authorities, housing
    authorities, and development commissions;
    and
2. private and local public activities or
    projects that need permits or licenses
    from state agencies.

Not included in MEPA coverage, however, were *projects of local city and town agencies* which neither
involve state funds nor require permits or licenses
from state agencies.
  The relationship between public and private responsibilities under MEPA was substantially qualified
by amendments to the law in 1974:

Any such agency, department, board, commission,
authority or any authority of any political
subdivision which grants permit determinations,
orders or other actions shall prepare an environmental impact report for any work, project, or activity of any private person, firm

112

or corporation which may cause damage to the environment and for which no funds of the commonwealth are to be expended, provided that such report shall be limited in scope to the subject matter jurisdiction of such agency, department, board, commission, authority or authority of a political subdivision by which said report is prepared. (Emphasis added)

Thus, while broad environmental review was still required for projects associated with the state, the review of private actions was limited in scope to the "subject matter jurisdiction" of the permitting agency. Rules and regulations issued in 1976 provided interpretation and definition on this matter. To facilitate the two different levels of review, the regulations established *two separate systems of environmental documentation;*[80] a public project with the potential for significant impact would require a preliminary Environmental Assessment Form (EAF) to determine whether or not draft and final Environmental Impact Reports (EIR's) would be necessary. A private project, on the other hand, would need only a Limited Environmental Assessment Form (LEAF) to determine the need for draft and final Limited Environmental Impact Reports (LEIR's). Take, for example, a private project needing a wetlands permit for a culverting action. The only impacts which would need to be considered in the environmental review would be those which are directly related to the culverting action, regardless of what other primary or secondary impacts may result.

The rules and regulations also established classes of categorically exempt actions. The most distinctive feature of this section was the inclusion of a "threshold exemption" category which allows agencies to establish *specific* thresholds of adverse environmental impact for their specific types of projects. Another more common provision was that of a blanket exemption for all "ministerial projects." This exemption applies to requests for permits or licenses "where there is not agency discretion involved in the granting or issuance thereof."

The responsibility of implementing and overseeing the state environmental review process fell primarily upon the Executive Office of Environmental Affairs (EOEA), a "super-agency," created through an overall reorganization of the state's environmental agencies. More specifically, it is the Environmental Impact Review Division, within the Secretary's Office, that coordinates and monitors the entire MEPA process; *all state environmental documents are submitted to the Division for mandatory review and comment.* Although the Secretary cannot order that an action be stopped on the basis of such review, he can request the Attorney General's Office to obtain a temporary restraining order to stop work until the law's obligations are met. In addition, the Division retains responsibility for the promulgation and revision of statewide rules and regulations for MEPA as well as the power to approve or disapprove the implementing procedures prepared by state agencies.

COST AND CASELOAD

Statewide Implementation

Between July 1973 (when MEPA became fully effective) and June 1977, there were 2,830 filings for projects with the Secretary of Environmental Affairs (this figure may be somewhat conservative since those agencies outside of the Executive Office are not required to file environmental assessment forms). Table 4.13 details the experience with preliminary environmental assessments and the preparation of draft EIR's for the last three years. *The trend is clearly toward the preparation of fewer documents;* as the initial backlog of projects is processed, the caseload tends to taper off and establish itself at a considerably lower level.[81]

TABLE 4.13
MASSACHUSETTS ENVIRONMENTAL
DOCUMENTATION*
1975-77

	FY 1975	FY 1976	FY 1977
Environmental Assessments	961	385	309
Draft EIR's (MEPA only)	27	19	15

*Categories include both full and limited documents.

Focusing on fiscal 1977, the ratio between public and private projects is readily discernible by breaking out the number of "limited" filings (see Table 4.14). Thus, the ratio of public to private actions requiring environmental documentation is approximately 4:1. This, of course, does not account for the many projects considered exempt under the law. *The intentional exclusion of locally permitted actions from consideration under the Act is probably responsible for the inflated ratio of public projects under review.*

114

TABLE 4.14		
EAF Versus EIR Preparation		
	# Documents	% Total
Environmental Assessments (EAF)	264	85%
Limited Environmental Assessments		
(LEAF)	45	15%
	309	100%
Draft EIR's	12	80%
Draft Limited EIR's	3	20%
	15	100%

Another interesting caseload characteristic is discovered by comparing the ratio of preliminary assessments to the preparation of impact reports. Such a comparison shows that *less than 5 percent of the total filings for 1977 required the preparation of an EIR.* A similar computation over the last three fiscal years yields a figure of between 3 and 6 percent. Thus, it is safe to generalize a *current ratio of approximately twenty EAF's for every EIR prepared (5 percent).*

Since a statewide assessment of MEPA costs has not been undertaken, we turn to an analysis of the costs incurred by the Coordinating Agency and several important construction agencies.

Executive Office of Environmental Affairs

As explained previously, responsibility for implementation of MEPA was assigned to the Secretary of Environmental Affairs. With the assistance of the Environmental Impact Review Division (MEPA Unit), he is responsible for the promulgation and review of general regulations and is required by law to comment on each EIR prepared.

The assessment of costs in this regard has been greatly simplified for two reasons:

1. Within the EOEA, there is one designated office for MEPA administration and review, and
2. The activities of this office are specifically accounted for in the daily statement of transactions.

The following series of cost breakdowns (Table 4.15) reflect the clearcut definition of responsibilities assigned to EOEA under MEPA.

115

TABLE 4.15

OFFICE OF ENVIRONMENTAL AFFAIRS

Secretary's Office and the Environmental Impact
Review Division[1]

Estimated Cost

	FY 75	FY 76	FY 77	FY 78
Salaries:				
Admin./Mgmt.	$11,338	$34,123	$44,123	$48,123
Review of EIR's	34,014	51,186	71,186	81,186
Subtotal	45,352	85,309	115,309[2]	129,309[3]
Consultant services for EIS review[4]	24,387	1,967		
Overhead	1,373	2,745		
Travel	548	600		
Office & Admin. Expenses	5,323	7,850		
Equipment & Supplies	--	--		
EQC Monitor	2,100	1,638		
Printing & Distribution	8,978	6,552		
Rental--Office space, xerox, etc.	3,355	2,850		
Subtotal	46,064	23,609	⌐20,000	⌐20,000
TOTAL Cost of Coordination	$91,416	$108,918	$135,309	$149,309

[1]The above cost estimates include:
- staff time for:
  a. review of incoming EIS's
  b. administration of the EIS program
  c. clerical workload
  d. review of pre-EIS environmental assessments
- supply and material expenses for publication and circulation of all guidelines.
- supply and material expenses for supplying miscellaneous information to EIS applicants.

The estimates do not include:
- litigation costs which are borne by the Attorney General's Office.
- public hearing costs which are borne by the developer or responsible agency.

[2]Increases are based on time estimates within EOA. Most other costs have remained constant. This figure reflects five new positions created in the second half of fiscal 1977. This raised the overall MEPA staffing from 4 to 9.

[3]Same as above except that increase reflects a full year of costs for the above-mentioned five positions.

[4]Consists entirely of assistance from The Institute
on Man and Environment at the University of Massachusetts.
Work has consisted primarily of implementation and control
procedures for handling the Act.

Department of Environmental Quality Engineering

DEQE is the Massachusetts environmental regulatory agency; in fact, as a result of reorganization
in 1975, DEQE assumed the regulatory responsibilities of the former Massachusetts Department of Natural Resources.

Being housed within the same executive office
as the Environmental Impact Review division does not,
however, alleviate DEQE of MEPA responsibilities.
As an initiator of state projects and a grantor of
permits and licenses, DEQE processes more material
through the "MEPA Unit" than any other Department in
the Executive Office of Environmental Affairs.[82]
(See Table 4.16)

In fiscal 1977, for example, DEQE prepared ninety environmental assessment forms (EAF's) for review. In the judgment of DEQE, seventy-six of these
EAF's warranted a negative classification (no potential for significant adverse impact), and fourteen a positive classification (potential for significant adverse impact). Of the seventy-six Negative EAF's submitted for review, two were overturned
by the Secretary of Environmental Affairs. The Secretary concurred with DEQE that all fourteen positive
EAF's warranted preparation of EIR's.

TABLE 4.16
DEQE CASELOAD (FY 1977)

	SUB-MITTED	NEG.	POS.	NEG. EAF'S APPROVED BY EOEA	NEG. EAF'S OVERTURNED BY EOEA	POS. EAF'S APPROVED BY EOEA
EAF'S	90	76	14	74	2	14
LEAF'S	39	36	3	34	2	3
TOTAL	129	112	17	108	4	17

DEQE submitted thirty-nine limited environmental
assessment forms (LEAF's) for private projects--thir-
ty-six as Negative LEAF's and three as Positive. Of
the thirty-six Negatives, two were overturned by the
Secretary while all three Positives were agreed to
require the preparation of LEIR's.

DEQE was required to produce a total of seventeen reports (EIR's and LEIR's) pursuant to the MEPA

process in fiscal 1977--fourteen for public and three for private sector projects. Nine of the fourteen public projects were statewide "208" water quality management reports. The remaining five were for a variety of projects averaging approximately $5,000 each for report preparation.[83] A figure of $5,000 is also considered to be a good estimate of the three private sector LEIR costs. *Thus, report preparation costs for DEQE pursuant to MEPA were in the neighborhood of $40,000 for fiscal 1977.*

The MEPA process within DEQE operates on an informal basis--DEQE does not make budget requests for funds to be appropriated directly for the environmental review process. One person within DEQE coordinates MEPA activities on a full-time basis. Other DEQE personnel put in lesser amounts of time, as required (primarily in the ad hoc review of documents). This averages out to a time commitment of approximately two person-years per year at a dollar figure of $30,000-$35,000.[84]

Therefore, the estimated total cost of the MEPA process in fiscal 1977 for DEQE was between $70,000-$75,000.

Department of Public Works

The Massachusetts Department of Public Works (DPW), like other transportation agencies, is heavily supported by Federal funding. Thus, both MEPA and NEPA compliance are necessary in the development of any project involving such funds. Since MEPA clearly defers to NEPA, however, the preparation of a single document satisfies both laws:

> Section 62G. In the case of projects for which an environmental impact statement is required under the National Environmental Policy Act of 1969, draft and final federal environmental impact statements may be submitted in lieu of environmental impact reports.

There are approximately 200 projects processed through DPW each year including those prepared by other agencies and reviewed by DPW. Out of this figure, four to five EIR/EIS's are prepared annually and eight to ten Federal Negative Declarations are used as state EIR's.[85] Thus, there are twelve to fifteen major Federal projects and ten to fifteen Federally funded actions of lesser significance each year. Of the remaining 170-175 filings, approximately 150 involve smaller state-wide projects.[86]

Another figure that has remained relatively stable over the last five years (save for inflation) has been the cost per mile of various project facets. These unit costs can be used to estimate the relative expense of environmental work:

$20,000/mi. - Administration/Management
$25,000/mi. - Document Preparation (Consultants)
$45,000/mi. - TOTAL

Construction costs (in 1978 dollars) have traditionally been $2.5 million per mile on the average. *Therefore, environmental analysis constitutes about 1.8 percent of construction costs.* Since many costs are consolidated by such analysis (i.e., a great deal of preliminary design is incorporated into the environmental work), these figures may, indeed, be somewhat inflated. We, therefore, turn to a breakdown of agency environmental costs (Table 4.17). The cost figures represent estimates for joint MEPA-NEPA compliance.

TABLE 4.17
DEPARTMENT OF PUBLIC WORKS
Estimated FY 1977 Cost

Environmental Section[1]	$ 240,000
11 professionals	
2 clericals	
District Personnel[2]	$ 120,000
8 professionals	
Division Personnel[3]	$ 15,000
Consultants[4]	$2,800,000
(EIS/EIR preparation)	
Overhead[5]	$ 375,000
(100% of salary)	
TOTAL[6]	$3,550,000

[1]Includes all personnel in the central office. Since the bulk of document preparation is done in the districts, however, this figure reflects mainly administrative and management costs. Out of the eleven professionals, two work specifically on MEPA related matters. Therefore, two people at $20,000 per year plus 100% overhead (see below) yields a separable central office MEPA cost of $80,000 per year.
[2]DPW has eight District Offices with one Environmental Engineer at an estimated $15,000 per year in each.
[3]An estimated one person-year to cover all environmental duties within the Headquarter's Divisions.
[4]The yearly average of contracts let to consultants for environmental studies.

119

[5]Includes travel, office space, equipment, materials, telephone, public information costs, indirect costs, and benefits.

[6]Total does not include the cost of litigation (which is purported to be minimal), hearings, or printing and distribution of documents.

Comparing the above figure of approximately $3.5 million per year for environmental work with the capital and operating budgets of the agency:

Operating Budget - $191 million
Capital Budget - $200 million

we see again that *environmental work constitutes about 1.75 percent of the capital costs and 1.8 percent of operations.*

Department of Public Utilities

No budget expenditure information was available for the Massachusetts Department of Public Utilities (DPU) since, unlike the Office of Environmental Affairs, no separate accountability of EIS review cost was made. Cost figures are, therefore, based on time estimates.

In fiscal 1975, a single Public Utilities Engineer spent 60 percent of his time working on drafting and implementing guidelines and 15 percent of his time in review and preparation of EIS's. The cost breakdown for 1975 was as follows:

Administration - $ 9,480
EIS Activity   2,370
Total $11,850

In fiscal 1976, this figure nearly doubled (see Table 4.18):

TABLE 4.18
DEPARTMENT OF PUBLIC UTILITIES
Estimated FY 1976 Cost

Public Utilities Engineer (1)	$ 9,075
Public Utilities Engineer (2)	7,475
Hearings	1,000
Travel Expenses	500
Printing	1,000
Supplying Miscellaneous Information to EIS Applicants	300
Total	$19,550

The above cost estimates include:
- staff time for the review of incoming EIS's.
- staff time for the preparation of EIS's.
- clerical workload incurred by the EIS program.
- staff time for review of pre-EIS environmental assessments.
- hearings cost for informal meetings specifically dealing with EIS issues and a small portion of formal hearing costs attributable to an EIS (formal hearings are geared to old licensing functions of Public Utilities Commission rather than the EIS process but discussion of EIS issues was included).
- travel expenses.
- expense of printing and circulating EIS's.
- cost of supplying miscellaneous information to EIS applicants.

The above cost estimates do not include:
- cost of resource inventories necessary for preparation and review of EIS's carried out by the DPU (an estimated $4,000-$5,000 per EIS).
- cost of other agencies for consulting functions.
- cost for review of Federal EIS's which is estimated at $1,500 for all Federal EIS's.
- staff time for administration of the EIS program.

Massachusetts Housing Finance Agency

The Massachusetts Housing Finance Agency (MHFA) does not employ staff specifically for environmental review and processing.[87] Analysis and review functions are performed by the Mortgage Department (usually one person with input from the Site Review Officer).

EIS's, when required, are prepared by the developer of a project or by a consulting firm. The cost of the EIS is borne by the developer, initially, and later included as part of the mortgage amount. In fiscal 1976, the agency submitted three EIS's to the Office of Environmental Affairs. The overall cost of these EIS's was obtained from MHFA files on the project:

EIS Cost
Boston East - 404 units in 2 stages   -Approx.$40,000
Boston East - 441 units outside Boston-Approx.$50,000
Riverbank Apt. - 114 units in Boston   -Approx.$ 9,000

The salary of the one staff person involved in the EIS process was $11,500 for fiscal 1976. This included the staff time for:

- review responsibility of incoming EIS's.
- the clerical workload.
- the research effort in review-site inspection.
- the review of pre-EIS environmental assessments.
- out-of-pocket travel expenses.

Costs not included but applicable to the EIS program include:

- the staff time for administering the review program; this involved the drafting of agency guidelines (which are almost a carbon copy of the guidelines issued by the Office of Environmental Affairs). So perhaps $1,500 could be added to the cost of the program for fiscal 1974.
- for supplying miscellaneous information to EIS applicants.
- for conducting a hearing on agency projects.

COST EFFECTIVE ADAPTATIONS

Although changes in the MEPA process have certainly proven significant since its inception in 1972, *recent amendments to the law constitute a virtual re-write of the original legislation.* The amendments, which respond to emerging problems of implementation, passed both the state Senate and House in December of 1977 and went into effect in mid-February of 1978. They contain both major revisions and refinements of existing language as well as several new provisions. Some of the more outstanding features include the following:

1. Scoping - A common problem cited by both the private and public sectors is the uncertainty associated with EIR substance and focus. This uncertainty results in the production of lengthy, encyclopedic documents that often fail to concentrate on key issues. The amendments respond to this problem by inserting a provision for early problem identification:

    If an (environmental imapct) report is required, the secretary with the cooperation of said person and agency shall, within the above-mentioned thirty day period limit the scope of the report to those issues which by

the nature and location of the
project are likely to cause damage
to the environment. The secretary
shall determine the form, content,
level of detail and alternatives
required for the report. In the
case of a permit application to an
agency from a private person for a
project for which financial assistance
is not sought the scope of said report
and alternatives considered therein
shall be limited to that part of the
project which is within the subject
matter jurisdiction of the permit.

Thus, the aim is to tailor each individual
EIR so that the process focuses only on
those issues of critical importance. This
technique of streamlining the environmental
review process on a case-by-case basis is
unique to Massachusetts although other
states (i.e., California and New York)
have attempted to focus the reporting process in a more programmatic way.[88] And
on the Federal level, CEQ is proposing a
similar technique in its draft rules and
regulations for NEPA implementation.

2. Timing - One of the most serious problems
associated with environmental review is the
delay associated with the processing and
review of documents--a problem which is of
especial concern to the private sector.
The amendments respond to this fundamental
problem by attempting to regularize and coordinate the notification, review and permitting process (see Figure 4.1). Wherever
possible, a specific time period has been
designated for the accomplishment of a
given task:

* Under the law, persons applying to an
  agency for a permit or for financial
  assistance must, within ten days after
  filing the first permit or assistance
  application, notify the secretary of
  environmental affairs of the nature of
  the project (an environmental notification (ENF) has replaced the environmental assessment form (EAF)).

FIGURE 4.1
NEW MEPA PROCESS

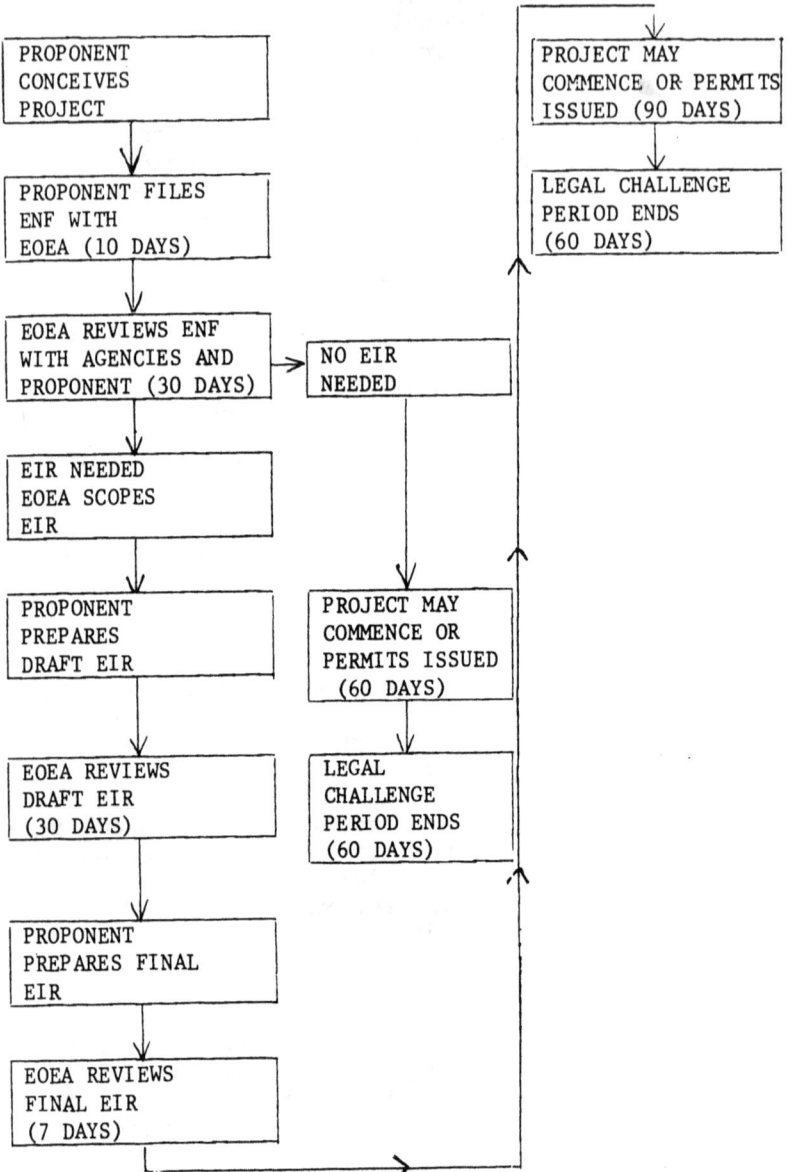

```
┌─────────────────────┐                          ┌─────────────────────────┐
│ PROPONENT           │                          │ PROJECT MAY             │
│ CONCEIVES           │                          │ COMMENCE OR PERMITS     │
│ PROJECT             │                          │ ISSUED (90 DAYS)        │
└─────────────────────┘                          └─────────────────────────┘
           │                                                  │
           ▼                                                  ▼
┌─────────────────────┐                          ┌─────────────────────────┐
│ PROPONENT FILES     │                          │ LEGAL CHALLENGE         │
│ ENF WITH            │                          │ PERIOD ENDS             │
│ EOEA (10 DAYS)      │                          │ (60 DAYS)               │
└─────────────────────┘                          └─────────────────────────┘
           │
           ▼
┌─────────────────────┐      ┌──────────────┐
│ EOEA REVIEWS ENF    │      │ NO EIR       │
│ WITH AGENCIES AND   │─────▶│ NEEDED       │
│ PROPONENT (30 DAYS) │      └──────────────┘
└─────────────────────┘
           │
           ▼
┌─────────────────────┐
│ EIR NEEDED          │
│ EOEA SCOPES         │
│ EIR                 │
└─────────────────────┘
           │
           ▼
┌─────────────────────┐      ┌──────────────┐
│ PROPONENT           │      │ PROJECT MAY  │
│ PREPARES            │      │ COMMENCE OR  │
│ DRAFT EIR           │      │ PERMITS ISSUED│
└─────────────────────┘      │ (60 DAYS)    │
           │                 └──────────────┘
           ▼                        │
┌─────────────────────┐             ▼
│ EOEA REVIEWS        │      ┌──────────────┐
│ DRAFT EIR           │      │ LEGAL        │
│ (30 DAYS)           │      │ CHALLENGE    │
└─────────────────────┘      │ PERIOD ENDS  │
           │                 │ (60 DAYS)    │
           ▼                 └──────────────┘
┌─────────────────────┐
│ PROPONENT           │
│ PREPARES FINAL      │
│ EIR                 │
└─────────────────────┘
           │
           ▼
┌─────────────────────┐
│ EOEA REVIEWS        │
│ FINAL EIR           │
│ (7 DAYS)            │
└─────────────────────┘
```

- Within thirty days of receiving the notice, the secretary is to decide whether
  an environmental impact report is required and establish the scope of the
  project report. The law directs the
  secretary to encourage preparation of
  the reports during the initial planning and design phase of projects.

- Agencies cannot undertake a project
  which is environmentally damaging or
  grant a permit or financial assistance
  until sixty days after the environmental
  report is made public.

- Upon availability of the draft or final
  environmental impact report, there will
  be a thirty-day public and agency review period, with a possible thirty day
  extension for "major and complicated
  projects."

- Within seven days after the public and
  agency review period, the secretary
  must state whether the report adequately
  complies with the law.

- Agencies must act on permit applications
  within ninety days after the final environmental impact report is made available, within ninety days after notice
  that a report is not required, or ninety
  days following a permit application,
  whichever is latest.

- For "major and complicated projects,"
  the environmental affairs secretary may
  establish specific procedures for evaluation and review of the project's environmental impacts.

By making the various steps of environmental review more predictable, it is felt the
cost of delay in the private sector will be
greatly diminished since plans can be based
on relatively firm commitments of time.[89]

Notification of the public concerning filings and decisions under these procedures
is accomplished through the bi-monthly publication of The Environmental Monitor which

is available free of charge from the Secretary of Environmental Affairs upon written request.

3. <u>Statute of Limitation</u> - This provision is an embellishment of the original statute of limitation established by the 1974 amendments. The most significant improvement over the original provision is the addition of a requirement to file a *notice of intention* to commence an action or proceeding pertaining to MEPA. Such a notice must be filed within sixty days of project clearance (or permit issuance in the case of private projects). This, it is hoped, will allow for a period of out-of-court problem solving before legal proceedings actually begin.[90] If problems are not resolved in this sixty-day period, then litigation must commence within thirty days for a private project and within sixty days for a public activity.

4. <u>Rules and Regulations</u> - The new amendments empower the secretary to issue regulations <u>binding</u> on all applicable agencies:

> Section 2. The secretary of environmental affairs shall after consultation with other secretaries of executive offices and with agencies not within executive offices, promulgate reasonable rules and regulations to carry out the purposes of sections 62 to 62H, inclusive.

New regulations were issued May 16, 1978, entitled "Regulations Governing Implementation of the Massachusetts Environmental Policy Act." With the amended MEPA, the two system procedure of EAF's and LEAF's is obsolete. Instead, there is one form, the Environmental Notification Form (ENF), which is filed by appropriate project proponents.[91]

The amendments also provide for the revision of both existing categorical exemptions and project categories that will <u>always</u> require environmental impact reports:

126

Section 62E. With the approval of the
secretary of the Executive office having
jurisdiction over an agency, or if an
agency is not within an executive office, with the approval of such agency,
the secretary of environmental affairs
shall establish general and special
categories of projects and permits
which shall or shall not require environmental impact reports based upon the
scope and duration of potential impacts
from the nature, size and location of
said projects or portions thereof which
require permits.

5. Coordination- Whereas environmental analysis involving the issuance of permits was
formerly conducted through the permitting
agency, the amendments now require all such
private actions to comply with MEPA through
the Secretary's Office:

Section 62C. Any environmental impact
report shall be submitted to the secretary of environmental affairs who shall
issue public notice of the availability
of such report.
A reviewing agency or person, and any
agency which has jurisdiction by law or
special expertise with respect to any
environmental impact involved may submit
written comments on any draft or final
environmental impact report to the secretary of environmental affairs who
shall affix any such comments which are
timely received to his statement on such
reports. Said reports and any comments
submitted in review thereof shall be
public documents.

The significance of this move toward *cen-
tralization* lies in recent court interpretations giving the secretary (and hence the
Environmental Impact Review Division) the
power to make threshold determinations for
EIR preparation. Given this ability, *the
secretary would possess the absolute power
to determine the need for impact report
preparation in both public and private sector projects.*

It will be interesting to see what effects the new amendments will have on the mix of costs and caseload. *Increased use of the negative assessment to achieve mitigative measures is already common practice in Massachusetts as it is in California and Washington.* The new provision for scoping, however, should provide an additional technique for balanced environmental consideration. If the offer of a negative assessment with attached mitigative measures is not well received by a project originator, the secretary could then require the preparation of an EIR focusing only on those issues which created the problem in the mitigative measures.[92] In this way, the proponent would be forced to deal with the root problems of a given project. *This technique marks a new direction in environmental assessment and its impact should be carefully monitored.*

Cost savings from the new amendments would be reaped primarily by private applicants and the permitting or construction agencies. Since EIR's would deal only with specified issues (identified through the "scoping" process), the cost of document preparation should be substantially reduced. This is especially important since *95 percent of all EIR's are written by consultants* despite encouragement by the rules and regulations for agencies to "use their own full-time staff as much as possible" in the preparation of reports. Permitting agencies would experience administrative cost savings since the amendments would route all EIR's requiring permits through the Secretary's Office. Hence responsibilities for MEPA coordination within the Environmental Impact Review Division would, if anything, be increased by the proposed amendments.

MINNESOTA

The Minnesota Environmental Policy Act, enacted by the state legislature in 1973, established a state environmental policy "to foster and promote the general welfare, to create and maintain conditions under which man and nature can exist in productive harmony, and fulfill the social, economic, and other requirements of present and future generations of the state's people."

To accomplish these goals, an EIS requirement was established which differed substantially from the NEPA model both in applicability and content:

Sec. 4. (116D.042) ENVIRONMENTAL IMPACT STATEMENTS. Subdivision 1. Where there is potential for significant environmental effects resulting from any major governmental action or from any major private action of more than local significance, such action shall be preceded by a detailed statement prepared by the responsible agency or, where no governmental permit is required, by the responsible person on:
(a) The environmental impact of the proposed action, including any pollution, impairment, or destruction of the air, water, land, or other natural resources located within the state;
(b) Any direct or indirect adverse environmental, economic, and employment effects that cannot be avoided should the proposal be implemented;
(c) Alternatives to the proposed action;
(d) The relationship between local short term uses of the environment and the maintenance and enhancement of long term productivity, including the environmental impact of predictable increased future development of an area because of the existence of a proposal, if approved;
(e) Any irreversible and irretrievable commitments of resources which would be involved in the proposed action should it be implemented;
(f) The impact on state government of any

federal controls associated with proposed action; and
(g) The multistate responsibilities associated with proposed actions.
(Emphasis added)

MEPA's applicability was limited, to some extent, by the necessity for governmental actions to be "major" and private actions to be "of more than local significance" before incurring the impact statement prerequisite. *Included under rubric of "private actions," however, are projects proposed to be undertaken by private persons that do not require a governmental permit.* Although it is doubtful that a private project of "more than local significance" will not also entail permitting at some level, such a qualification did represent an attempt to better define those actions likely to require an EIS. The inclusion, in the EIS, of such considerations as the impacts of Federal controls on state government (subsection (f)) and the multistate responsibilities associated with the proposed action also represented significant extensions of traditional EIS content.

Responsibility for administering the environmental impact statement system was assigned to the Environmental Quality Council (EQC), an interagency body composed of the heads of seven state agencies (State Planning, Pollution Control, Natural Resources, Energy, Agriculture, Health, and Highways); a representative of the Governor's Office, and four citizens selected from the Citizen's Advisory Committee (CAC). Staff for the EQC is located within the State Planning Agency.

In addition to assigning the Council a wide range of responsibilities, the legislation also vested the EQC with significant authority:

Subd. 2. The Minnesota environmental quality council shall, by January 1, 1974, prescribe by rule and regulation in conformity with provisions of Minnesota Statutes, Chapter 15, guidelines and regulations setting forth those instances in which environmental impact statements are required to be prepared for new and existing actions, including the time and manner in which such statements shall be prepared and acted upon, and to coordinate the processing of such statements among local, state, and federal agencies. The council may require the preparation of an environmental impact statement for any action or project not referred to

> in its guidelines and regulations. Further,
> the council may require the revision of an en-
> vironmental impact statement which is found to
> be inadequate. (Emphasis added)

*A significant feature unique to the Minnesota law is
that the EQC is authorized to reverse or modify a
proposed action if it determines that the action
would be inconsistent with the declaration of policy
contained in the Act.*

MEPA also outlined mechanisms for citizen involvement throughout the EIS process. Included were
provisions for early public notice of natural resource management and development permit applications as well as requirements that environmental
documents be available for public scrutiny. Probably of greatest significance, however, was the provision for *citizen petition* as a means of initiating
environmental review:

> Subd. 3. Upon the filing with the council
> of a petition of not less than 500 persons
> requesting an environmental impact state-
> ment on a particular action, the council
> shall review the petition and, where there
> is material evidence of the need for an en-
> vironmental review, require the preparation
> of an environmental impact statement in ac-
> cordance with provisions of this section.

Furthermore, public meetings or hearings are required as part of the draft EIS review process. However,
formal public hearings on EIS's alone are a rarity,
primarily due to cost; they are held only when required for permit issuance under another statute.
Informal public meetings are preferred because the
responsible department does not have to follow formal hearing procedures.[93]

Rules and regulations were first issued by the
Council in April 1974. Under the regulations, the
EQC held a tight rein on the implementation of the
Act. More specifically, EQC had the authority to
require preparation of an EIS (threshold determination); to designate an agency or person to be responsible for the preparation of the document (lead
agency); and to review EIS's for adequacy (enforcement authority).

Proposed projects were initially brought to the
attention of the EQC by a state agency responsible
for granting permits, by a developer requesting determination of whether an EIS is required, or by a

citizen petition containing 500 signatures. Before
a case could actually be heard by the Council at one
of its monthly meetings, however, it first had to be
considered by the "EQC Technical Committee." The
Technical Committee, composed of representatives of
each of the agencies on the EQC, met once a week to
receive technical information and citizen input on
individual projects. It was the Technical Committee's responsibility to do a detailed analysis of the
project and to prepare a report and recommendation to
the Council on what action should be taken. The EQC
then voted either to follow the Technical Committee's
recommendation or to take some alternative action.

In a study of the "Implementation of the Minnesota Environmental Policy Act"[94] commissioned by
EQC in 1974, this centralized implementation strate-
gy was evaluated as ineffective and cumbersome:

> The Environmental Quality Council has become
> so overburdened with the hearing of individual
> action proposals for an EIS determination that
> it has had increasingly little time to direct
> its attention to the matters which were in-
> tended to be its primary foci: issues of state-
> wide environmental policy, interagency coordin-
> ation, administration of the critical areas
> program, and power plant siting. Further,
> the Council is decreasingly able to give ade-
> quate attention to each case individually.
> Something must be done to re-orient the Coun-
> cil's work schedule immediately.[95]

In a subsequent effort to decentralize the process, amendments to the rules and regulations were
submitted in 1975. The amendments (which were finally approved in February 1977) passed several former EQC responsibilities onto state agencies and local units of government. Public agencies are now
responsible for developing their own departmental
rules, regulations, and guidelines for noticing and
acting upon certain permits and subsequent environmental documents while the Council's primary EIS review function involves the *hearing of appeals* (see
Figure 4.2). In essence,the amended regulations
passed the responsibility of *threshold determination*
from EQC to the EIS responsible agencies. *The EQC
makes the decision as to the need for an EIS only if
valid objections are made to the sponsoring or initiating agency's decision. If an EIS is prepared, however, the Council must review the final EIS to ensure its adequacy.* Final decisions to approve or

FIGURE 4.2
MINNESOTA ENVIRONMENTAL REVIEW PROCESS
DECENTRALIZED
(effective Feb. 1977)

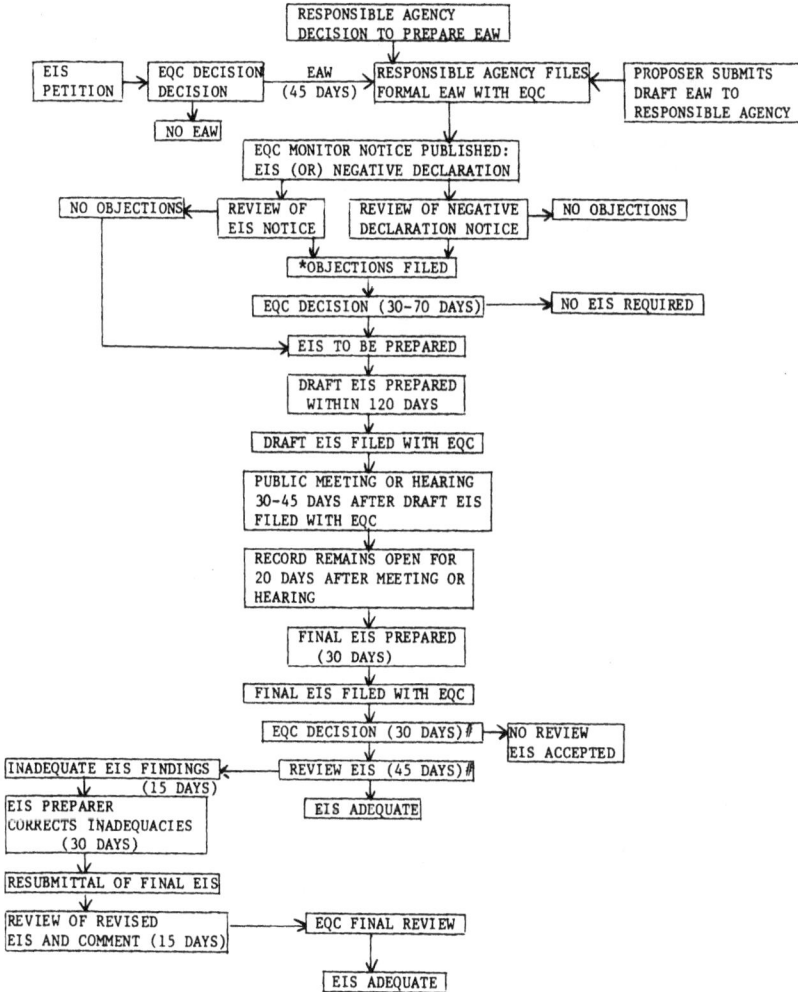

```
                        ┌─────────────────────────┐
                        │ RESPONSIBLE AGENCY       │
                        │ DECISION TO PREPARE EAW  │
                        └─────────────────────────┘
┌──────────┐  ┌──────────────┐   EAW    ┌─────────────────────────┐  ┌──────────────────┐
│ EIS      │  │ EQC DECISION │          │ RESPONSIBLE AGENCY FILES │  │ PROPOSER SUBMITS │
│ PETITION │  │ DECISION     │(45 DAYS) │ FORMAL EAW WITH EQC      │  │ DRAFT EAW TO     │
└──────────┘  └──────────────┘          └─────────────────────────┘  │ RESPONSIBLE      │
                     │                                                │ AGENCY           │
                ┌─────────┐                                           └──────────────────┘
                │ NO EAW  │
                └─────────┘
                     ┌──────────────────────────────────┐
                     │ EQC MONITOR NOTICE PUBLISHED:     │
                     │ EIS (OR) NEGATIVE DECLARATION     │
                     └──────────────────────────────────┘
┌───────────────┐  ┌────────────┐   ┌──────────────────────┐  ┌───────────────┐
│ NO OBJECTIONS │  │ REVIEW OF  │   │ REVIEW OF NEGATIVE   │  │ NO OBJECTIONS │
│               │  │ EIS NOTICE │   │ DECLARATION NOTICE   │  │               │
└───────────────┘  └────────────┘   └──────────────────────┘  └───────────────┘
                        ┌──────────────────┐
                        │ *OBJECTIONS FILED │
                        └──────────────────┘
                     ┌──────────────────────────┐   ┌───────────────────┐
                     │ EQC DECISION (30-70 DAYS) │──▶│ NO EIS REQUIRED   │
                     └──────────────────────────┘   └───────────────────┘
                        ┌────────────────────┐
                        │ EIS TO BE PREPARED │
                        └────────────────────┘
                        ┌────────────────────┐
                        │ DRAFT EIS PREPARED │
                        │ WITHIN 120 DAYS    │
                        └────────────────────┘
                        ┌─────────────────────────┐
                        │ DRAFT EIS FILED WITH EQC │
                        └─────────────────────────┘
                        ┌─────────────────────────┐
                        │ PUBLIC MEETING OR HEARING│
                        │ 30-45 DAYS AFTER DRAFT EIS│
                        │ FILED WITH EQC           │
                        └─────────────────────────┘
                        ┌─────────────────────────┐
                        │ RECORD REMAINS OPEN FOR  │
                        │ 20 DAYS AFTER MEETING OR │
                        │ HEARING                  │
                        └─────────────────────────┘
                        ┌─────────────────────┐
                        │ FINAL EIS PREPARED  │
                        │ (30 DAYS)           │
                        └─────────────────────┘
                        ┌─────────────────────────┐
                        │ FINAL EIS FILED WITH EQC │
                        └─────────────────────────┘
                        ┌────────────────────────┐   ┌───────────────┐
                        │ EQC DECISION (30 DAYS)# │──▶│ NO REVIEW     │
                        └────────────────────────┘   │ EIS ACCEPTED  │
                                                      └───────────────┘
┌──────────────────────────┐  ┌─────────────────────┐
│ INADEQUATE EIS FINDINGS  │◀─│ REVIEW EIS (45 DAYS)#│
│           (15 DAYS)      │  └─────────────────────┘
└──────────────────────────┘        ┌───────────────┐
┌──────────────────────┐            │ EIS ADEQUATE  │
│ EIS PREPARER         │            └───────────────┘
│ CORRECTS INADEQUACIES│
│ (30 DAYS)            │
└──────────────────────┘
┌──────────────────────────┐
│ RESUBMITTAL OF FINAL EIS │
└──────────────────────────┘
┌──────────────────────────┐   ┌──────────────────┐
│ REVIEW OF REVISED        │──▶│ EQC FINAL REVIEW │
│ EIS AND COMMENT (15 DAYS)│   └──────────────────┘
└──────────────────────────┘       ┌───────────────┐
                                    │ EIS ADEQUATE  │
                                    └───────────────┘
```

Review time variable dependent upon meeting or hearing schedule
* From pertinent agencies, petition or developer

commence a project cannot be made until after this environmental review process is complete.

Statewide Implementation

The 1977 Legislature directed the Minnesota Environmental Quality Council (EQC) to compare the costs of the current system of preparing EIS's with the costs of establishing a separate office to prepare the documents. The study involved surveys of most agencies that have prepared EIS's, examination of several environmental documents, and discussions with numerous technical experts. The report, entitled "Environmental Impact Statement Preparation: Is a Central Office the Answer?"[96] and issued February 15, 1978, arrived at two major conclusions:

1. Establishing a separate office to prepare EIS's entirely in-house is not recommended. The current system, which involves many state and local agencies and several options for document preparation, is very flexible. This is an important factor, since EIS workload is extremely variable. It also enhances cooperation among agencies and increases the environmental awareness of agencies and project proposers.
2. The cost to the state of the current system is low--a rough estimate is $2.5 million for the first forty-two EIS's prepared by state and local agencies. There are two major reasons for this: first, private developers have provided a great deal of background data and also paid for much of the EIS preparation costs; and, second, state trust funds and federal moneys have been involved in many EIS's. The 1977 MEPA regulations requiring most private proposers to pay set amounts toward EIS preparation costs should further reduce the cost to the state.

Since 1974, when the original environmental review program rules were issued, several public and private projects have been reviewed. As of December 1, 1977, these included:

- 88 citizens' petitions
- 156 Environmental Assessments (the Environ-

mental Assessment (EA) was the preliminary
screening document under the original rules).
- 65 Environmental Assessment Worksheets (EAW)
(the new screening document).
- 60 Environmental Impact Statements.
- Numerous other projects were determined not
to meet mandatory review categories or otherwise require review.

Of the sixty EIS's ordered, eight were later rescind-
ed when the project was dropped or substantially altered. *State agencies have prepared very few of the
preliminary documents--only 9 percent of the EA's
and 11 percent of the EAW's. However, state agencies are responsible for almost two-thirds of the
EIS's.* Most of the documents have been prepared by
the Minnesota Department of Natural Resources (DNR),
the Minnesota Pollution Control Agency (PCA) and the
Minnesota Department of Transportation (Mn/DOT).
Federal agencies have not prepared any EA's or EAW's,
but are responsible for four EIS's.
Local units of government have prepared almost
half of the EA's, and 81 percent of the EAW's, but
they have prepared less than one-third of the EIS's.
Eighty-eight different local units have been sponsoring or initiating agencies. Only a few local units
have prepared three or more preliminary documents.
No local unit has prepared more than one EIS. *These
numbers indicate that most projects in the State of
Minnesota do not require EIS's.* In contrast, approximately 3,800 Environmental Impact Reports were
prepared under California's Environmental Quality
Act in 1974 alone.[97]
Since most agencies do not keep detailed records of EIS costs, it was not possible to determine
the exact costs of the current system of environmental document preparation. Indeed, document preparation is so often intermixed with other related
activities, including preliminary design discussions,
site reviews, permit application, and policy development, that determining exact costs would have required considerable research.
Time constraints prevented researching the cost
of individual EIS's. The determination of the *state
costs* of EIS preparation was based on the agencies'
detailed estimate for eight EIS's studied in depth,
agencies' rough estimate of twenty-three EIS's, and,
for the remaining EIS's, the costs of an EIS similar in project type, amount of agency effort involved and preparation method. When review costs
and costs of miscellaneous assistance from state

TABLE 4.19
TOTAL EIS PREPARATION COSTS TO STATE, 1974-1977
(42 EIS's Received by December 1, 1977)

Project	Responsible Agency Costs	State Agency Review, Misc. Help Costs
In-House preparation *8 highway EISs	$0 (paid by federal funds and state trust funds)	24,000
Mankato Beltline	25,000	3,000
Edenvale, Preserve PUDs	0 (federal funds)	6,000
*Saganaga Lodge	20,128	933
*Lower St. Crois Master Plan	10,000	3,000
*Williams Marina	20,128	933
MMCD	62,800	3,000
Calhoun-Isles Resid.	20,000	3,000
*Minnesota Zoo	291,000	3,000
Responsible Agency consultant (excludes amount paid by proposer)		
*MP&L Clay-Boswell	131,334	3,506
*NSP Sherco	131,334	3,506
*Reserve Mining	1,400,000	3,000
*Shakopee Bypass	0 (state trust funds)	3,000
Opus II	3,093	1,823
Wooddale	53,313	2,418
3M Office Park	2,118	1,671
Red Rock Asphalt Plant	2,118	1,671
Jonathon New Town	0 (federal funds)	3,000
Tenth St., St. Cloud	116,000	3,000
Review of proposer-supplied information analysis		
*Inland Steel Mining	3,821	2,376
*Oglebay-Norton Mining	50,000	3,000
*Williams Pipeline	3,821	2,376
*Northern Pipeline	3,821	2,376
Boise-Cascade Bridge	2,118	1,671
Breezy Point	2,118	1,671
Sherwood Forest	2,118	1,671
*NSP-TR-1	3,366	1,737
*3 other power lines	10,098	5,211
*CU-TR-1	3,821	2,376
Other preparation method		
Winona Flood Control	10,000	3,000
Cedar-Riverside	20,128	933
*MP&L (Warroad)	10,000	3,000
	2,423,596	104,859

*State Responsible Agency   2,528,455 TOTAL

agencies were not available, $3,000 per EIS was in-
cluded.

Based on the above techniques, the total EIS
preparation costs to major state agencies for the
forty-two EIS's received by the EQC from 1974 to
December 1, 1977, were estimated to be roughly
$2,500,000 (see Table 4.19). Of this, $330,000 was
incurred by local agencies for thirteen EIS's. These
figures are based on estimates from key staff; most
likely, the figures underestimate supervisory and
miscellaneous staff time, office materials, staff
fringe benefits and travel costs.[98]

With this broad overview of statewide EIS costs
in mind, we now turn to a brief examination of key
agencies in the environmental arena.[99]

The Environmental Quality Council

The central role of coordination and review
played by the EQC in MEPA implementation has been
described in detail in the previous section. The
costs incurred by the Council in carrying out its
duties can be computed by:

1. Using time estimates, salaries, and other
    estimates of expended costs; or
2. Proportioning salary requests with appro-
    priation.

Beginning with the first technique (Table 4.20), it
is noticed that the cost of environmental review increased substantially from 1975 to 1976. The figures in Table 4.20 estimate the total cost to EQC
as $120,290 in fiscal 1975 and $93,985 in fiscal
1976. These figures are probably an underestimation
because the hearing cost and EQC Monitor publication
costs are sizable costs (see appropriations in next
section) which are not included. Also, when adding
in the borrowed staff time of Power Plant Siting and
the EQC Administrator for the fiscal 1976, the cost
estimate could rise $10,000-$12,000. *Therefore, a
more probable range of costs is $130,000-$135,000
for fiscal 1975 and 1976.* Using the second technique,
that of balancing out the budget request with the
legislative appropriations, the breakdown of expenditure estimates for fiscal 1976 is as follows:

TABLE 4.20
ENVIRONMENTAL QUALITY COUNCIL

STAFF (salaries reflect range for position)	Estimated Costs FY75		FY76	
	Admin.	EIS Activity	Admin.	EIS Activity
Manager			7,582	11,373
Planner III	3,000	17,700	3,123	17,700
Planner II		18,513		18,513
Planner III		4,164		1,041
Planner II (75) and Planner III (76)		962		1,041
Intern		1,200		
Secretary				6,440
Intern				1,680
Intern				1,680
Research Scientist I				2,800
Environmental Prog.Mgr. & State Plng. Dir.	12,637		8,400	
Dir., State Planning Agency	6,000		3,000	
EQC Chairman	3,450		6,900	
Subtotals	25,087	42,503	29,005	62,280
Fiscal Year Totals	67,590		91,285	

The above figures reflect the cost of:
- reviewing all incoming state EIS's.
- reviewing all state environmental assessments.
- administering the EIS program--drawing up the rules and regulations.
- participating on the interagency task forces, EQC and Technical Committee.
- all associated clerical work.
- coordination of review on Federal EIS's.
- testifying at hearings pursuant to MEPA.

The above figures do not reflect the cost of:
- the Power Plant Siting Section of EQC that coordinates the EIS's pertaining to that subject matter (could possibly add $8,000 for 2 planners putting in 20% of their time to matters directly related to the EIS).
- travel expenses.
- conducting hearings on EQC rules and regulations; appeals.
- other SEPA related work (e.g., publishing annual report).
- legal expenses for hearings.
- publication of EQC Monitor.

The above figures do not reflect the following costs but additional estimates were given for them:

	FY 75	FY 76
1. Contractor time for a special review related to MEPA's implementation.	$50,000	--
2. Supplies and material expenses including publication and circulation of guidelines	$ 1,200	$1,200
3. Capital outlay--equipment.	$ 1,500	$1,500

Salaries:

	Manager	$23,158
	Planner III	19,072
	Planner II	13,554
	Planner II	16,925
	Planner III	18,078
	Intern	8,276
Total		$99,063

Legal...................$12,103
   (a) hearings on EAW appeals
   (b) hoping to hire a special assistant
      to the Attorney General
EQC Monitor.............$14,905
   (a) expect their cost to increase to $10,000
      for the fiscal 1976
Program Review..........$ 9,937
   (a) monitoring the effectiveness of the
      rules and regulations
EIS Chargeback..........$ 2,484
   (a) developing a fee structure (not yet
      started)

TOTAL..................$138,492

Thus, the two estimates are very closely aligned and suggest a cost to EQC in the neighborhood of $130,000-$140,000 per year.

Department of Natural Resources

The role played by DNR in the EIS process is primarily one of EIS *preparation and review.* The costs associated with these functions can also be computed by:

1. Using time estimates, salaries, and other estimates as given by the staff of DNR; or
2. Proportioning the budget request with legislative appropriations.

Using the first technique, it is estimated that DNR spent $39,557 in fiscal 1975 and $122,294 in fiscal 1976 for staff time for review of environmental documents, for preparation of EIS's, for providing review, advice and assistance to state and local agencies and private individuals in the preparation of plans, proposals, environmental assessments, and EIS, and for staff time on interagency task forces such as the Technical Committee and EQC. In fiscal 1975, $5,000 and in fiscal 1976, $8,700 was spent on travel, office supplies, and material expenses for producing, printing, and circulating DEIS's and FIES's. Approximately $3,356 was spent in fiscal 1975, and approximately $6,400 in fiscal 1976 for the administration of the EIS program by the Bureau of Planning and Protection (BEPP). This figure represents the cost of bringing statutory authority, administrative regulations and current policies into conformity with the objectives of the Minnesota Environmental Policy Act. *Therefore, the estimated total expenditures within DNR for the EIS Program are:*[100]

> FY 1975 - $ 47,913
> FY 1976 - $137,394

The above cost estimates do not include:

- the staff time for review of other state permits and programs.
- costs for supplying miscellaneous information to EIS applicants.
- staff time for review of NEPA documents.
- public hearing costs (DNR has held none pursuant to an EIS project).

Using the second technique, that of proportioning budget requests with actual appropriations, it was observed that the Department of Natural Resources was appropriated the sum of $169,000 for the 1976-78 biennium. These funds were to be spent on the following activities:

- Preparation of mandatory categories and procedures by agencies to determine when an Environmental Clearance Worksheet (CEW) would be completed and the method to be followed.
- Preparation of ECW.
- Notice of Negative Declaration.
- Notice of EIS preparation.
- Preparation of EIS.
- Review of petitions, CEW's, Notices of Negative Declaration, EIS preparation notices, and EIS.
- Preparation of materials for hearings.
- Review of "Findings" of hearings.

DNR allocated the fiscal 1976 portion of the $169,000 to be spent in the following manner:

	Total Salary
Bureau of Environmental Planning and Protection	
Planner 3, State	$15,340
Planner 1, State	10,130
Senior Clerk Typist	7,860
Division of Minerals	
Planner 3, State	15,340
Division of Waters	
Senior Hydrologist	17,190
Hydrologist 1	12,140
Total Salaries	$78,000
Office Equipment	1,200
Printing	2,000
Travel	2,400
TOTAL	$83,600

The above total does not include all the factors listed under Technique I, such as:

- staff time of the director (salary of $13,621) of BEPP who works full time on state EIS activity, but is paid from another source.
- borrowed staff from other divisions within the department that spend time on EIS activity. The cost of this borrowed staff amounts to about $31,443.
- borrowed staff that would aid in bringing the statutory authority, administrative regulations, and current policies into conformity with objectives of MEPA which amounts to about $4,423.

Including the above factors, the estimated cost using Technique II is $133,087. *Therefore, balancing the two approaches, a probable range of the cost to DNR for the EIS Program is $130,000-$140,000 for fiscal 1976.*

The Pollution Control Agency

The role played by the PCA in the EIS process is also one of document *preparation and review.* The costs associated with these functions can be computed by:

1. Using time estimates, salaries, and other estimates of expended costs; and

2. Proportioning salary request with appropriation.

Estimates for fiscal 1976 using the first technique are as follows in Table 4.21.

TABLE 4.21
THE POLLUTION CONTROL AGENCY
Estimated FY 1976 Cost

	Admin-Cost	EIS Activity
Office of Environmental Analysis		
Coordination	2,376	3,596
Secretary	4,560	
Coordination Water Quality		9,375
Secretary		3,600
Coordination Air Quality		8,225
Economist Research Sci. I	5,100	5,100
Chief, Water Pollution Control		
- the Tech. Rep.	15,400	6,600
	27,436	36,916
TOTAL	$64,352	

The above figures reflect the cost of:
- reviewing all incoming state EIS's.
- reviewing all state EA's.
- preparing EIS's.
- administering the EIS program.
- participating on the interagency task forces--EQC and
  Technical Committee.
- all associated clerical workload.
- testifying at some hearings pursuant to Minnesota Environmental Policy Act.

The above figures do not reflect the cost of:
- testifying at additional hearings spent pursuant to MEPA.
- resource inventories that went into the preparation and
  review of environmental documents.
- the time that the regional staff spent on project documents review on Federal EIS's.

The above figures do not reflect the following costs but additional estimates were given for them:
- contractor time for the preparation of
  (a) the Clay Boswell EIS-Power Plant Siting $350,000
  (b) the NSP Sherco EIS-Power Plant Siting   450,000
- travel expenses, communications and postage   3,600
- soft supplies   3,000
- printing   4,000
- equipment for new staff   1,500
- hearings on agency projects (includes $200

142

per day for court reporter, $35-$100
per day for expert testimony, $30 per day
for incidental expenses which approxi-   8,850
mates $295 per day, no room rental)
- a training course on EIS implementation   1,000

The total cost estimates of expenditures (not including the contractor cost) is, therefore $86,302.
This is probably a slight overestimation because:

1. estimates were always raised as opposed to
    lowered when talking among individuals for
    the time estimates;
2. salaries quoted are higher than that listed
    on the budget request;
3. the projects quoted for hearing costs were
    not prepared by the PCA so they would not
    have incurred all the costs themselves.

The second technique used was balancing budget
requests and legislative appropriations. By making
allowance for staff changes and the difference between EIS budget requests and EIS budget appropriation, the cost breakdown is as follows:

Salaries
   (a) Agency Coordinator $16,330
   (b) Assistant Coordinator
       for Air, Noise &
       Solid Waste   2,482
   (c) Assistant Coordinator
       for Water Quality   9,562
   (d) Senior Clerk Typist   4,560
Subtotal $32,888 $32,888
Benefits   4,277
Equipment   3,465
Services   1,485
Travel   4,882

TOTAL $46,997

Also computed on the budget request was an additional $17,480 in PCA commitment for borrowed staff with-
in the agency. *Thus, total EIS commitment was
$64,477.* As is evident by the computations in Tech-
nique I, PCA underestimated time commitment. *By balancing the two means of calculation, PCA's cost in
the state EIS program was $76,000-$82,000 for fiscal
1976.*

COST EFFECTIVE ADAPTATIONS

The rules and regulations governing MEPA were substantially revised in 1977 to improve program effectiveness and also reduce costs to Project-Originating Agencies, Review Agencies, and project sponsors. Some important aspects in this regard include:[101]

1. *The replacement of the narrative Environmental Assessment (EA) with the EAW, a short worksheet requesting specific environmental information.* This reduces the amount of unnecessary information and analysis included and also minimizes "packaging" costs. Also, the EAW may be skipped if the sponsoring agency and the proposer agree an EIS is needed. *The changed procedure for EAW review also reduces costs.* Before, the EQC and its advisory Technical Committee reviewed each EA at meetings in St. Paul and determined whether an EIS would be needed. The two or more trips to St. Paul for often day-long meetings for the local officials, project proposer, and other interested persons are no longer required. Meetings of the Technical Committee and EQC are considerably shorter, releasing these people for other duties. Less EQC staff time is required to handle administrative matters, saving about twenty hours per project.

2. *The revised rules encourage environmental document preparation early in the project approval process,* when many design decisions remain to be made; environmental protection measures can be discovered and incorporated easily. Dollar savings for this are unavailable, but studies on housing costs indicate the substantial increase in costs caused by delay in environmental review procedures.

3. *Specific categories of projects are set forth* (e.g., any industrial, commercial, or residential development of forty or more acres) which, by definition, either require or are exempt from EAW preparation. General areas of exemption (e.g., when a substantial portion of an action has been completed) are also outlined in the regulations--conspicuously missing is any mention

of "ministerial actions" although individual agencies are encouraged to develop
their own system of exemptions.

4. *The new rules and regulations contain provisions for charging back the costs of EIS
preparation to private project applicants*
in accord with a formula based on the cost
of the project. The maximum required payment is .3 percent of the project cost between $1-$10 million, .2 percent of the project cost between $10-$50 million, and .1
percent of the project cost over $50 mil-
lion; this would be $132,000 for a $75 million project and $37,000 for a $15 million
project. There is no chargeback for projects under $1 million which typically in-
volve small amounts of land and little construction. The "chargeback rules" provide
for cooperation between the project proposer and the permitting agency, who together agree on what information is needed
in the EIS and which party will provide it.
The EQC becomes involved only if there is
a dispute. The proposer gets up to a one-
third credit for information he provides
that is used in the EIS; this can be increased if the sponsoring agency agrees.

Finally, the environmental impact statement process is well coordinated with the regulatory and permitting process. This is accomplished through an
Environmental Permit Coordination Program which provides a "master application" option to all applicants
requiring more than one permit from the state. Furthermore, the coordination of land-use planning with
the environmental review process is encouraged
through the provision for a "Related Actions EIS,"
in which an environmentally sound comprehensive plan
is submitted as an EIS.

NEW YORK

New York State's Environmental Quality Review
Act (SEQR), the most recently enacted SEPA law, was
passed on August 1, 1975. As such, New York was
able to take considerable advantage of the experience
gained in the implementation of NEPA as well as other
state-level EIS programs in drafting its law. The
Act outlined a broad basis for its implementation:

> §8-0101. Purpose. It is the purpose of this
> act to declare a state policy which will en-
> courage productive and enjoyable harmony be-
> tween man and his environment; to promote ef-
> forts which will prevent or eliminate damage
> to the environment and enhance human and com-
> munity resources; and to enrich the under-
> standing of the ecological systems, natural,
> human and community resources important to
> the people of the state.

The impact statement prerequisite, which went
into effect on June 1, 1976, was imposed on any "action" which may have a "significant effect on the
environment." What constituted an "action" was
straightforwardly defined as including:

> (i) Projects or activities directly under-
> taken by any agency; or supported in whole
> or part through contracts, grants, subsidies,
> or involving the issuance to a person of a
> lease, permit, license, certificate or other
> entitlement for use or permission to act by
> one or more agencies;
> (ii) Policy, regulations, and procedure-making.

Conversely, "actions" were defined as not including:

> (i) Enforcement proceedings or the exercise
> of prosecutorial discretion in determining
> whether or not to institute such proceedings;
> (ii) Official acts of a ministerial nature,
> involving no exercise of discretion;

146

(iii) Maintenance or repair involving no substantial changes in existing structure or facility.

Thus, New York followed the lead of California by
adopting the somewhat arbitrary exemption of "ministerial actions;" it distinguished itself, however,
by *legislatively mandating that applicability of the
law be comprehensively extended to local as well as
private actions;* most other SEPA statutes had looked
to legal interpretation or the adoption of rules and
regulations to clarify this point.
Furthermore, the description of environmental
impact statement content, although modeled after
that of NEPA and ensuing state-equivalents, added
several important entries to the list of requirements:

> 2. All agencies shall prepare, or cause to
> be prepared by contract or otherwise an en-
> vironmental impact statement on any action
> they propose or approve which may have a
> significant effect on the environment. Such
> a statement shall include a detailed state-
> ment setting forth the following:
> (a) A description of the proposed action and
>   its environmental setting;
> (b) The environmental impact of the proposed
>   action including short-term and long-term
>   effects;
> (c) Any adverse environmental effects which
>   cannot be avoided should the proposal be
>   implemented;
> (d) Alternatives to the proposed action;
> (e) Any irreversible and irretrievable
>   commitments of resources which would be
>   involved in the proposed action should
>   it be implemented;
> (f) Mitigation measures proposed to minimize
>   the environmental impact;
> (g) The growth-inducing aspects of the pro-
>   posed action;
> (h) Effects of the proposed action on the use
>   and conservation of energy resources; and
> (i) Such other information consistent with
>   the purposes of this article as may be
>   prescribed in guidelines issued by the
>   commissioner pursuant to §8-0113.
> (Emphasis added)

The addition of the "mitigation measures" and "growth

147

inducing aspects" entries drew on the experience of
already existing state legislation in California.
The "energy conservation" and "additional information" entries, however, represented *new concepts in
the legislative definition of EIS content.*

It is interesting to note that subdivision 2
(above) <u>explicitly</u> gives agencies the right to contract out the preparation of EIS's. The law also
allows agencies to charge a fee to an applicant in
order to recover the costs incurred in preparing an
impact statement. In order to avoid duplication,
environmental reporting under SEQR is fully coordinated with and made in conjunction with the Federal
requirements under NEPA.

The law establishes a predictable system of
timing for the environmental review process. After
the filing of a draft EIS, a public hearing (if one
is deemed necessary) must begin within sixty days
of the filing and, unless the proposed action is
withdrawn, the final impact statement must be prepared within four to five days after the close of
the hearing. The establishment of criteria for
whether or not a proposed action may have a significant effect on the environment was left to be clarified by the mandated preparation of rules and regulations.

The responsibility for the promulgation of
statewide rules and regulations for the implementation of SEQR fell upon the Department of Environmental Conservation (DEC) and more specifically, the
<u>Office of Environmental Analysis</u> (OEA). Although
SEQR provided no special review authority for DEC
vis a vis other agencies, it did encourage the review of those projects for which DEC may possess
special expertise. Other state laws (involving regulatory permit programs) also facilitate DEC as a
party to decisions on regulatory issues and major
resource questions.[102] Thus, the major roles of the
Office of Environmental Analysis with respect to
SEQR are implementation, coordination, and discretionary review.

The initial Statewide Rules and Regulations
(6 N.Y.C.R.R., Part 617) issued in March of 1976,
clarified many areas of uncertainty concerning SEQR
implementation. They encouraged individual agencies
to develop their own criteria for environmental review, but insisted that agency implementing regulations be no less protective of the environment than
the statewide regulations. The fee for recovering
the costs of preparing environmental impact statements for applicants was set at a ceiling of *one-half*

of one percent of the total project cost. This figure was arrived at through empirical cost studies (discussed earlier in this chapter) in California comparing the relative costs of environmental documentation to overall project costs.[103]

The Rules and Regulations also established the specifics of EIS content and defined the procedures for review. The most significant area of clarification, however, concerned the issues of *threshold determination* and *exemption*. After outlining ten specific criteria for determining significant environmental impact, a list of actions containing critical thresholds was devised according to the following classification system:

> Type I   Actions or classes of actions that are likely to require preparation of environmental impact statements because they will in almost every instance have a significant effect on the environment.

> Type II   Actions or classes of actions which have been determined not to have a significant effect on the environment and which do not require environmental impact statements under this Part.

(The full descriptions of Types I and II are presented in Appendices B7-B9.)

The purpose of this section was to simplify the task of determining whether or not a proposed action may be significant and thus require more complete environmental analysis. "Type II" actions are very much reminiscent of the "Categorical Exemptions": developed in other states while the designation of actions likely to require an EIS--the "Type I" category--is a technique much less frequently used.

As originally enacted, SEQR would have been effective for both state and local agencies on June 1, 1976. However, the fact that implementing procedures, as defined by OEA, were necessarily general and statewide in orientation meant that little specific guidance could be offered to the many localities required to implement the law. Although localities were encouraged to create procedures more specific to their own concerns, a lack of experience in this area combined with political questions of state authority vs. "home rule" resulted in low levels of local compliance with SEQR. As a result, the law was amended in June of 1976 to delay mandatory implemen-

tation at the local level. The controversial "Phased Implementation" amendment established the following timetable:

> After 9/1/76, all state agencies directly under-
> taking an action which may have a significant
> environmental effect must prepare an EIS.
> After 6/1/77, the requirement is extended to
> local agency actions and actions subsidized
> by state agency grants.
> After 9/1/77, the requirement is extended to
> include actions requiring permits or li-
> censes from state and local agencies.

The idea was to allow localities more time and leeway in establishing their programs since the required mixture of expertise becomes increasingly difficult to assemble at lower levels of government.
    Before this three-phase plan would be fully ex-
pedited, however, the law was again amended (1977)
pushing back and altering the phasing schedule still
further. This time, the designation of "Type I" action established in the original rules and regulations was introduced as a criteria in the phasing:

> June 1, 1977 - Type I actions (likely to re-
> quire an environmental impact statement)
> which are directly undertaken by local gov-
> ernments, and Type I actions which are fund-
> ed by state agencies are subject to the law's
> requirements.
> September 1, 1977 - Type I actions requiring a
> license or permit from state or local govern-
> ment and Type I actions which are funded by
> local government will be subject to the law's
> requirements.
> September 1, 1978 - All other actions.

*Thus, only large-scale projects would be subject to
the requirement for environmental impact statement
prior to September 1978.* The amendments also delayed implementation at the local level, but did require local governments to adopt necessary procedures
prior to the effective dates.
    State and local agencies were also directed to
compile lists of projects, within their jurisdictions, which had been approved and were therefore,
not subject to the environmental impact statement
requirement. Lists of these *grandfathered actions*
were to be submitted to the appropriate chief fiscal
officer who would certify that "substantial time,

work, or money have been expended" on the projects
prior to the appropriate SEQR implementation deadline. This provision responded, by and large, to
the serious problems encountered under NEPA and in
California with respect to retroactive application
of the EIS requirement to on-going projects. Although the amendments established a definite timetable for the submission of "grandfather" project
lists, the criteria for determining what constitutes
an "approved" or "committed" project was somewhat
more elusive.

COST AND CASELOAD

Statewide Implementation

Because it is new, SEQR offers a unique chance
to trace the budget changes caused by the enactment
of an impact statement program. It is possible to
look at not only requested and actual appropriations
for the implementation of SEQR, but also how this
budgetary information compares to actual time commitments (i.e., "borrowed" or "unbudgeted" time).
By assessing the trade-offs between these two types
of cost estimates (budgeted vs. actual time commitments), a picture of how EIS program costs become
absorbed or *sublimated* within already existing activities begins to develop. The "phasing" aspect
of SEQR should also afford the opportunity to trace
relative cost impacts of state, local, and private
implementation.
Since only "major" or "Type I" actions are included under the law at this point, the bulk of the
analysis will deal with state level costs. The implications of local level implementation, however,
were assessed for a local assistance program in fiscal 1976-77 and are currently the topic of a study
being conducted by the Syracuse-Onondaga County
Planning Agency (SOCPA).
Some 1,612 units of local government in the
state are affected by the SEQR law (62 counties,
62 cities, 931 towns and 557 villages). With over
1,600 units affected, it is not unreasonable to estimate the preparation of 1,200 or more *impact
statements* per year. Using only the bare minimum
costs (as suggested in the "local assistance program" projection) of $2,000 per EIS and a figure of
1,200 statements annually, costs to local government units for the preparation of impact statements
under SEQR will be at least $2,400,000 annually.[104]
This figure represents preparation costs only and

does not include costs for: required public notices, agency costs for review of commentary on impact statements, agency costs for public hearings and hearing records, mailing costs, costs for reproduction of multiple copies of reports, etc. It is estimated that these costs will add another 30 percent of the minimum figure cited above, increasing to approximately $3,000,000 the total minimum cost to be added to local government expenditures annually as a result of preparation and administration of impact statements.[105]

The foregoing estimates deal only with projects and actions proposed by localities which result in actual preparation of impact statements. SEQR requires that all actions be assessed for potential environmental effects. In response to this requirement, processes and procedures will be installed in local government administration for preparation of abbreviated *environmental assessments* used to determine whether or not an impact statement will be required. Based on experience, less than 10 percent of actions which may have some potential for environmental effect actually result in the necessity for a full impact statement. It is estimated, therefore, that some 12,000 actions may be handled annually by local government, which will require some degree of environmental analysis, but not full impact statements. At a minimum, it is expected that environmental analysis on each individual action may generate a $200 extra cost. Based on that figure, and the 12,000 estimated above, this process will result in an additional $2,400,000 expenditure by local governments annually.[106]

At minimum, therefore, it is estimated that the SEQR law, when fully implemented, will result in additional costs to local government of:

$3,000,000 (Impact Statement Preparation)
$2,400,000 (General Review Process)
$5,400,000 TOTAL MINIMUM ANNUALLY

Keeping these local level cost estimates in mind, we now examine the components of state-level implementation.

Department of Environmental Conservation

As an agency subject to SEQR, DEC must institute its own environmental review system in accordance with the promulgated rules and regulations and (by virtue of its regulatory powers) prepare environmental impact statements on Department actions which may significantly affect the environment. Indeed, Table 4.22 shows DEC to be a major originator of actions requiring environmental documentation.

TABLE 4.22 NEW YORK STATE CASELOAD* under SEQR (as of Dec. 1, 1977)				
	Negative Declarations	Notices of Intent	Draft EIS's	Final EIS's
Dept. of Environmental Conservation	98	9	7	1
Other State Agencies	188	1	5	4
Local Agencies	113	10	8	-
*EIS figures do not include the preparation of 27 programmatic impact statements.				

The role of the Department in the administration of the law includes:

1. Receipt and review of draft impact statements.
2. Providing comments to agencies concerning the environmental impact of proposed actions and the adequacy of impact statements.
3. Receipt of final impact statement and filing for public availability.
4. Continued monitoring of the program to ensure effectiveness and compliance.
5. Issuance of determinations as to which agency is to be the "Lead Agency" for actions involving more than one agency.
6. Continued assistance and advice to agencies and the public concerning execution of the program, including technical services on a fee basis as required.

The bulk of these coordinating and review responsibilities were delegated to the Office of Environmental Analysis (OEA) within DEC.

1. Office of Environmental Analysis

OEA was formulated through a reorganization of existing units and activities in order to provide a coordinated system for analyzing the environmental impact of state, private, and Federally-funded projects. The function of OEA can be broken down into four major areas:

- Assisting Federal agencies in assessing the impact of proposed projects within New York State pursuant to NEPA.
- Conducting energy reviews including power plant siting, transmission line location, and radiation impacts.
- Various permitting duties; and
- Implementing and administering SEQR.

The following breakdowns display the relative size of the SEQR budget as compared to OEA as a whole for the last three fiscal years.

<table>
<tr><td colspan="3" align="center">TABLE 4.23
SEQR BUDGETARY BREAKDOWNS</td></tr>
<tr><td></td><td align="center">Office of
Environmental Analysis</td><td align="center">SEQR PROGRAM
(new funding)</td></tr>
<tr><td>FY 1976-77</td><td></td><td></td></tr>
<tr><td>Requested</td><td>$1,598,560</td><td>$400,000</td></tr>
<tr><td>Appropriated</td><td>874,466</td><td>-0-</td></tr>
<tr><td>FY 1977-78</td><td></td><td></td></tr>
<tr><td>Requested</td><td>$2,160,069</td><td>$210,710</td></tr>
<tr><td>Appropriated</td><td>1,481,187</td><td>132,791</td></tr>
<tr><td>FY 1978-79</td><td></td><td></td></tr>
<tr><td>Requested</td><td>$1,750,936</td><td>$ 3,600</td></tr>
<tr><td>Appropriated*</td><td>NA</td><td>NA</td></tr>
</table>

*Appropriations have not yet been made for fiscal 1978-79.

The $400,000 request in fiscal 1976-77 represented the first effort at getting the SEQR program funded since the law did not become effective until June 1, 1976. Therefore, a breakdown of this request should afford a good (although probably somewhat inflated) estimate of the additional burden placed upon OEA by SEQR:

OFFICE OF ENVIRONMENTAL ANALYSIS
Requested Appropriations for SEQR
FY 1976-77

CENTRAL OFFICE[1]

1 Principal Environ. Analyst ($21,545)	$ 21,545
3 Assoc. Environ. Analysts ($17,424)	52,272
1 Senior Environ. Analyst ($15,684)	15,684
1 Environ. Analyst ($10,714)	10,714
1 Stenographer ($6,450)	6,450
1 Clerk/Typist ($5,871)	5,871
SUBTOTAL	$112,536

```
REGIONAL OFFICES[2]
   9 Assoc. Environ. Analysts($17,424)        $156,861
   6 Environ. Analysts($10,714)                 64,284
   7 Clerk/Typists($5,871)                       41,097
   SUBTOTAL                                    $262,242

Total Salaries                                 $374,778
Travel                                           14,000
Equipment                                        11,200

TOTAL BUDGETARY REQUEST                         $399,978
------------------------------------------------------------
INDIRECT COSTS[3]
   Benefits (35% salary)                       $131,172
   Overhead (35% salary + benefits)             177,083

TOTAL COST                                      $708,233
```

[1]Added personnel here would form a "SEQR Unit" within OEA. This unit would provide information and program direction to regional environmental analysis units, develop and provide for a continuing information system for the benefit of agencies and the public concerning SEQR through appropriate communication media, provide day to day advice as required to agencies concerning execution of their responsibilities under the SEQR law, maintain records and provide reports on program effectiveness, review and coordinate review of impact statements prepared by State agencies, make recommendations concerning needs for revisions in rules and regulations and statutory provisions, arrange for appropriate review of impact statements by Department program units in Albany as required and coordinate a system for consultation and review services including a fee system.

[2]There are nine regional offices of DEC. Thus, this personnel addition would place one Associate Environmental Analyst in each of the regions with appropriate support.

[3]The indirect costs are calculated separately by the Office of Fiscal Management. Since they amount to considerable additional costs, they are included here for reference.

Thus, a liberal or "upper bound" estimate for the costs of SEQR administration and coordination is in the neighborhood of $700,000 per year. A closer look at Table 4.23, however, reveals that no money was actually appropriated for this function in fiscal 1976-77. *Responsibilities under SEQR must, therefore, have been met through the "borrowing" of already existing staff* (i.e., by adding new duties to already funded positions).

A brief analysis of the 1977-78 SEQR budget request confirms this assumption; it estimated SEQR manpower requirements at 33,050 person-hours. Available manpower (i.e., funded) was set at 1,700 person-hours leaving 31,350 person-hours unfilled. Additional staffing requirements were thus requested as follows:

EIS Review	6,000 hrs.
Local Requests	4,800 hrs.
Public Requests	1,600 hrs.
Doc. Preparation	1,800 hrs.
Clerical	14,850 hrs.
TOTAL	29,050 hrs.

This translated into an estimated cost of $198,179 plus a request for equipment, materials, and supplies of $12,531 for a total budget request of $210,710. Virtually all of the personnel would be located in the regions:

CENTRAL OFFICE	
Assoc. Environ. Analyst	$ 17,429
Typist	5,871
Air Quality Analyst (Air Resources)	21,545
REGIONAL OFFICES	
Assoc. Environ. Analyst (Region 2)	17,429
Sr. Environ. Analyst (Regions 1,4,5,6&9)	67,020
Environ. Analyst (Region 3)	10,714
Sr. Clerk (Regions 2,3,8&9)	28,816
Typist (Regions 1,4,5,6&7)	29,355
TOTAL	$198,179

Actual appropriations, however, included only $132,791 for the SEQR program. As a result, all requested positions in the central office were eliminated, leaving only ten newly funded positions in the regions. By adding indirect costs and estimates of legal counsel and hearing costs to the appropriation figure, we obtain a more accurate picture of the overall budgeted SEQR costs for fiscal 1977-78 as shown in Table 4.24. Thus, only $122,966 was appropriated for new positions. Based on the original 29,050 person-hour staffing requirement, this leaves 13,325 person-hours unfilled resulting in at least this much staff "borrowing" in order to fulfill SEQR responsibilities.

Indeed, it was possible to estimate the actual amount of time spent on SEQR by OEA Central Office

```
                        TABLE 4.24
                   BUDGETED SEQR COSTS
                    Estimated 1977-78

APPROPRIATIONS
  New Staff (10 Regional Positions)            $122,966
  Equipment, Materials, Supplies                  9,825
  Subtotal                                     $132,791
LEGAL COUNSEL¹                                    40,000
HEARINGS²                                          9,500
INDIRECT COSTS
  Benefits (35% salary)                          57,038
  Overhead (35% salary + benefits)              77,000

TOTAL                                          $316,329
```

¹This figure represents staff costs for DEC lawyers only.
One attorney is also retained outside of DEC, part-time, for
writing rules and regulations.
²Based on 19 hearings conducted thus far for SEQR at an
average cost of $500 per hearing. The expense is actually
incurred by the Office of Hearing Officers.

staff.[107] The Central Office, it must be remembered,
has not been given any additional funding for carrying out its SEQR responsibilities. Hence, any time
spent on SEQR is *unbudgeted* and *borrowed* from existing program staff:

OFFICE OF ENVIRONMENTAL ANALYSIS
Central Office
"Borrowed" Staff Time for SEQR Implementation

	Grade	Salary	SEQR Expense
Principal Environmental Analyst	27	$26,908	$26,908
Environ. Analyst	14	$10,714	$10,714
Sr. Landscape Architect (½ time)	23	$21,933	$10,967
Assoc. Environ. Analyst	23	$21,933	$21,933
Assoc. Environ. Analyst (min. time)	23	$18,052	-0-
Environ. Analyst	14	$11,679	$11,679
2 Clericals at $6,450/yr.	5	$12,900	$12,900
TOTAL			$95,101

*Therefore, it is currently necessary for the Central
Office to "borrow" nearly $100,000 worth of time*

each year in order to fulfill its minimum responsi-
bilities under SEQR. Looking back to the requested
appropriations for fiscal 1976-77, we see that this
figure nearly duplicates the $112,536 request for
Central Office personnel--the "SEQR Unit" that was
never funded. This suggests three possible conclusions:

1. There was already adequate staff to absorb
    the SEQR duties with no appreciable real-
    location of responsibilities. This would
    imply that OEA has been overstaffed.
2. That SEQR duties are accomplished through
    the addition of responsibility to existing
    full-time positions. This would imply a
    significant burden on the current staff.
3. That SEQR duties are accomplished through
    reprioritizing or "borrowing" existing
    staff time. This would imply that, al-
    though all of the mandated functions of OEA
    are being performed, none of them can be
    performed as completely as they should be.

The actual situation, no doubt, is a combination of
the three. *In any case, it is clear that the costs
incurred by the Office of Environmental Analysis in
coordinating SEQR range somewhere between $316,329
and $708,233 annually.*
    Finally, the internalization or "sublimation"
of costs outlined in the above may be further demonstrated by a quick look at the proposed 1978-79
budgetary requests for the SEQR program:

SEQR Informational Bulletins printing and mailing (57¢/copy including mailing x 1600 copies x 3 bulletins)	$2,700
Mailing of Revised SEQR Handbooks on request (1,000 copies at 90¢/copy)	900
Total	$3,600

Thus, we see how quickly the visible, budgetary out-
lays for an environmental impact statement program
may be absorbed into the general operating procedures of the Coordinating Agency. In this case, the
SEQR program may be costing in excess of $700,000
per year to administer at the state level while ear-
marked budgetary requests total only $3,600.

The New York State Department of Transportation (DOT), like other state transportation agencies, incurs both NEPA and SEQR related costs in its construction activities as a result of the Federal-aid program. The Department prides itself on its decentralized structure. It has divided its operation into ten regions plus the New York City area. Virtually all environmental analysis takes place within the regions. The main office intervenes only when there is insufficient staff in the regions to handle a given problem.[108]

The Main Office is also highly decentralized. The "Environmental Analysis Section" functions mainly as an informational unit. It keeps abreast of new policy developments so as to assist the regions in their applied environmental work.[109] Staff of the "environmental section" consists of individuals with the following expertise:

Number of Staff	Area of Expertise
2 people	air quality
2 people	noise abatement
1 person	socio-economic impact
1 person	cultural resources
1 person	administration/management

Expertise in the areas of water quality, landscaping, and land-use were not included in the Central Office when it was formed. Instead, these have remained within their original units. The Planning Division also incurs environmental costs through its "location study" function. *Thus, the costs associated with NEPA and SEQR are extremely diversified, making them difficult to assess.*

Adding to the problem of assessing the costs of the environmental impact statement process within DOT is the use of the "double-entry" accounting system in its budgetary process. In this system, cost items are recorded both by project and function. Thus, cost information can be discerned for a specific project as well as for general agency functions. Unfortunately, neither the "project" nor the "function" entries present a detailed enumeration of the costs of environmental review. For example, the costs of the EIS process are subsummed under the heading of "Program Planning and Project Management" which includes virtually every agency planning function prior to the commencement of construction.[110] Also, assessing environmental review costs on a pro-

ject basis would require the analysis of every proposed project on an individual basis.

Therefore, due to various circumstances peculiar
to New York State DOT, no reliable cost data could
be generated. Using the order-of-magnitude figure
of 2 percent operating budget derived in other state
transportation department analyses (California,
Washington, Massachusetts), however, we can approximate the costs associated with NEPA/SEQR compliance.
In fiscal 1975-76, DOT had an operating budget of
$220,767,200 while in fiscal 1976-77, they requested
$333,699,000. Using a rounded operating figure of
$300,000,000 per year, then a reasonable cost esti-
mate of the environmental review process is:

$300 million x .02 = $6 million per year

*One trend noted at DOT was that the caseload
has been steadily dropping off since the outset of
NEPA.* It began in the area of fifteen to sixteen
EIS's prepared each year and now averages five to
six impact statements per year.[111] The principal
explanation given for this reduction was the gradual
elimination of backlogged projects held up because
of the grandfathering issue. Indeed, once out of
this "transition period," the costs associated with
this diminishing caseload probably tend to decrease
as well.

Office of Parks and Recreation

The New York State Office of Parks and Recreation (OPR) manages the public recreational resources
of New York State. These resources include 145 major park facilities spread throughout about one-
quarter million acres of public lands which are located outside the Catskill and Adirondack State
Parks. OPR divides the state into eleven regions
and operates on an annual budget of $57 million.[112]

Existing statutes require that documentation
for the review of possible environmental effects be
part of any application or request for action made
to OPR. Most of these applications involve Federal
Bureau of Outdoor Recreation (BOR) grant-in-aid program requests making compliance with NEPA as well
as SEQR a necessity. The Office also initiates projects of its own, making it subject to the requirements of SEQR and/or NEPA (depending on funding
source). The approach taken to these environmental
review requirements reflects the mixture of regulatory and project-originating responsibilities peculiar to OPR.

160

Responsibilities under these statutes has been assigned to the Environmental Management Bureau (EMB)--a centralized unit within the Main Office. EMB retains a staff of eight professionals responsible for the processing, review, and disposition of all proposed actions. The Bureau is also responsible for the promulgation of agency implementing regulations under SEQR.

The most recent edition of these regulations (January 1977) contain several interesting approaches which differ from those proposed by DEC. For example, no environmental clearance can be given until all permits and licenses are secured by applicable state or Federal agencies. In fact, Negative Declarations are denied if any such regulatory approval is not secured.[113] Clearance from EMB, however, assures compliance with both NEPA and SEQR.

A fundamental difference between the implementation of SEQR by OPR and that suggested by DEC is that *no projects are considered categorically exempt from environmental review.* More specifically, all actions require environmental assessment; this does not mean that all actions will necessarily require the preparation of a SEQR EIS. Classification type (Figure 4.3) determines the level of environmental documentation that will be required for each action. Thus, every action requires at least the preparation of the initial "Memorandum of Review" in order for a Negative Declaration to be issued.

We now focus more closely on the costs associated with the two types of projects processed through EMB--grant-in-aid applicants and agency-initiated actions.

1. The Federal Grant-In-Aid Process

The Federal Bureau of Outdoor Recreation (BOR) has delegated both program and NEPA compliance responsibilities to OPR. In response to this, EMB has established the following procedure:

7. Environmental Clearance Process--in general, this process involves the following steps:
(a) receipt of EMB of the environmental documentation prepared for an eligible proposal;
(b) review of the eligible proposal by EMB;
(c) if necessary, EMB acts as a consultant to the applicant in order to assist in the development of a formal Environmental Assessment

161

Figure 4.3
NYS PARKS AND RECREATION
CLASSIFICATION PROCESSES

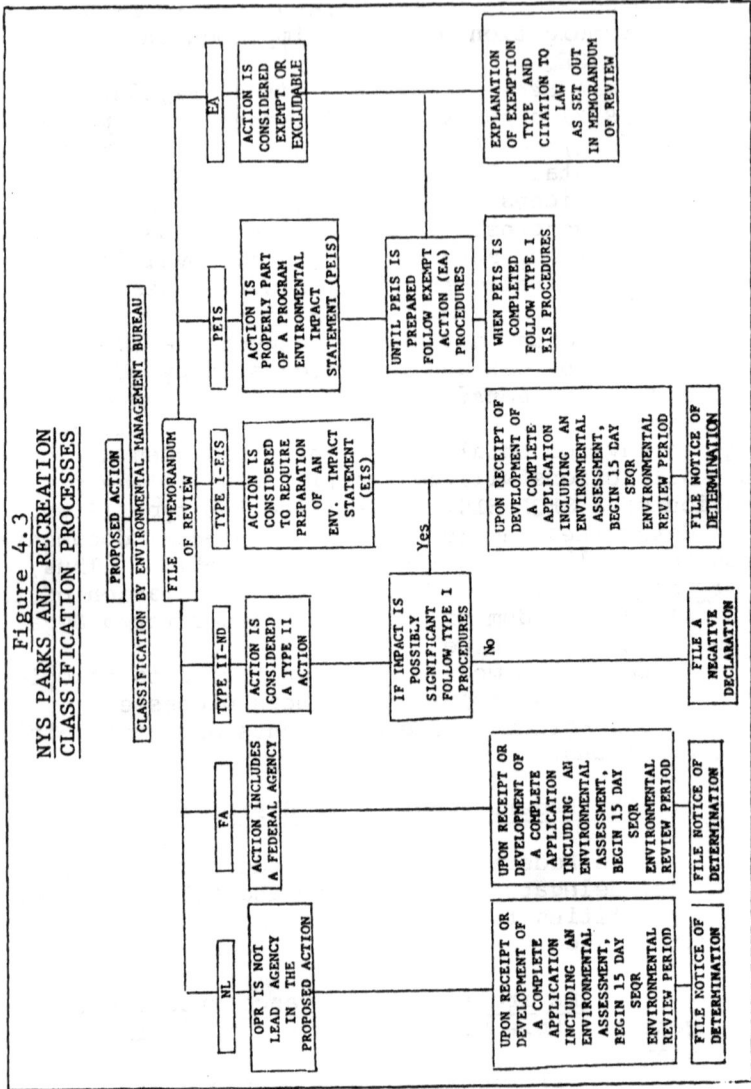

PROPOSED ACTION

CLASSIFICATION BY ENVIRONMENTAL MANAGEMENT BUREAU

FILE MEMORANDUM OF REVIEW

NL — OPR IS NOT LEAD AGENCY IN THE PROPOSED ACTION

UPON RECEIPT OR DEVELOPMENT OF A COMPLETE APPLICATION INCLUDING AN ENVIRONMENTAL ASSESSMENT, BEGIN 15 DAY SEQR ENVIRONMENTAL REVIEW PERIOD

FILE NOTICE OF DETERMINATION

FA — ACTION INCLUDES A FEDERAL AGENCY

UPON RECEIPT OR DEVELOPMENT OF A COMPLETE APPLICATION INCLUDING AN ENVIRONMENTAL ASSESSMENT, BEGIN 15 DAY SEQR ENVIRONMENTAL REVIEW PERIOD

FILE NOTICE OF DETERMINATION

TYPE II-ND — ACTION IS CONSIDERED A TYPE II ACTION

IF IMPACT IS POSSIBLY SIGNIFICANT FOLLOW TYPE I PROCEDURES

No

FILE A NEGATIVE DECLARATION

Yes

TYPE I-EIS — ACTION IS CONSIDERED TO REQUIRE PREPARATION OF AN ENV. IMPACT STATEMENT (EIS)

UPON RECEIPT OF DEVELOPMENT OF A COMPLETE APPLICATION INCLUDING AN ENVIRONMENTAL ASSESSMENT, BEGIN 15 DAY SEQR ENVIRONMENTAL REVIEW PERIOD

FILE NOTICE OF DETERMINATION

PEIS — ACTION IS PROPERLY PART OF A PROGRAM ENVIRONMENTAL IMPACT STATEMENT (PEIS)

UNTIL PEIS IS PREPARED FOLLOW EXEMPT ACTION (EA) PROCEDURES

WHEN PEIS IS COMPLETED FOLLOW TYPE I EIS PROCEDURES

EA — ACTION IS CONSIDERED EXEMPT OR EXCLUDABLE

EXPLANATION OF EXEMPTION TYPE AND CITATION TO LAW AS SET OUT IN MEMORANDUM OF REVIEW

for the eligible proposal;
    (d) EMB classifies the Eligible Pro-
posal either as Type I, which will require the
preparation of an Environmental Impact Statement or as Type II. A Type II determination
requires the preparation, filing, circulation,
and publication of a Negative Declaration. This
determination is backed up by the Environmental
Assessment prepared and cleared for the proposal;
    (e) Once the Type I or Type II deter-
mination has been published by the EMB as required by statute, the Eligible Proposal is <u>now</u>
a candidate for consideration for a BOR Grant-
in-Aid. It is a violation of both State and
Federal statutes, rules and regulations to se-
lect a candidate without first clearing the environmental, historic preservation and various
regulatory procedures.[114]

For the period September 1, 1976 to December 1, 1977,
EMB reviewed 224 such eligible proposals at a value
of approximately $74 million. Of these, 155 were
cleared, representing $47 million worth of project
work. The costs of this process may be broken down
as follows:[115]

1. Pre-Application Work - 2.6 staff years
    (processing, review and tentative classifi-
    cation)

    2.6 staff years = 572 staff days (based on
        220 staff days per year)
    572 staff days ÷ 224 eligible proposals =
        <u>2.55</u> staff days per pre-application
        review

2. <u>Clearance Work</u> - 1.9 staff years
    (review of revised environmental documenta-
    tion and determination)

    1.9 staff years = 418.5 staff days
    418.5 staff days ÷ 155 classified pro-
        posals = <u>2.70</u> staff days per clearance
        review

Thus, a total of 5.25 staff days was invested in
every eligible proposal. Assuming an average staff
salary of $23,029 per year, this amounts to about
$550 spent on each proposal for environmental review.
Multiplying times the 224 proposals received, *EMB
spent about $123,200 per year reviewing Federal*

163

grant-in-aid proposals. (This estimate does <u>not</u> include the costs of benefits, overhead, or legal counsel.)

2. Agency-Initiated Actions

During 1977, some 234 state projects were reviewed for SEQR and/or NEPA compliance amounting to $16.5 million worth of work. Seventy of these projects required assessments, thirteen of which also required full EIS's. Although no figures were available concerning departmental cost of preparation and review of the assessments, it was estimated[116] that the EIS's averaged $6,700 to prepare. The cost of EIS preparation for agency actions in 1977 was, therefore, approximately $87,100.

COST EFFECTIVE ADAPTATIONS

Aside from the issues concerning phased implementation and grandfathering, several other points were altered or added by the March 1977 amendments to SEQR. Throughout the legislation, the tone of environmental consideration was changed from one of *advocacy* to one of *compromise.* For example, subdivision 9 of the "Legislative findings and declarations" section was altered to read as follows:

> 9. It is the intent of the legislature that all agencies which regulate activities of individuals,corporations, and public agencies which are found to affect the quality of the environment shall regulate such activities so that (major) due consideration is given to preventing environmental damage.

Another important change was the addition of an EIS "focusing" clause similar to the one in California:

> Such statement (EIS) should be clearly written in a concise manner capable of being read and understood by the public, should deal with the specific significant environmental impacts which can be reasonably anticipated and should not contain more detail than is appropriate considering the nature and magnitude of the proposed action and the significance of its potential impacts.

This desire for shorter, more concise EIS's was supplemented by an amendatory clause stressing the need for full coordination between the impact statement process and other regulatory procedures:

> Notwithstanding the specified time periods established by this article, an agency shall vary the times so established herein for preparation, review and public hearings to coordinate the environmental review process with other procedures relating to review and approval of an action. An application for a permit or authorization for an action upon which a draft environmental impact statement is determined to be required shall not be complete until such draft statement has been filed and accepted by the agency as satisfactory with respect to scope, content and adequacy for purposes of paragraph four of this section. Commencing upon such acceptance, the environmental impact statement process shall run concurrently with other procedures relating to the review and approval of the actions so long as reasonable time is provided for preparation, review and public hearings with respect to the draft environmental impact statement.

In response to the many changes brought about by the amendments, revised statewide regulations were promulgated and adopted January 24, 1978. A significant portion of the regulations addressed the *grandfathering issue* described above:

> 617.12. Excluded actions.
> (a) When an action is an excluded action, the action and any subsequent agency approvals relating thereto are not subject to the requirements of this Part.
> (b) When an action is not excluded, but such action has received an approval by an agency prior to the effective dates of SEQR, such agency approval is not subject to the requirements of this Part.
> (c) An action is excluded if it has met any one · of the following requirements:
>   (1) in the case of a directly undertaken state agency action, the director of the budget has certified on or before September 1, 1976, that substantial time, work and money were expended on the action prior to June 1, 1976;
>   (2) in the case of a directly undertaken local agency action, the chief fiscal officer of the local agency has certified on or before

September 1, 1977, that substantial time, work or money were expended on the action prior to June 1, 1977;

(3) in the case of a state agency funded action, the chief fiscal officer of each such state agency which is funding the action has certified on or before October 1, 1977 that the application for funding was approved or in an approvable form prior to September 1, 1977;

(4) in the case of a local agency funded action, the chief fiscal officer of each such local agency which is funding the action has certified on or before December 1, 1977 that the application for funding was approved or in an approvable form prior to November 1, 1977;

(5) in the case of an action involving a subdivision of land, the action has received prior to September 1, 1977 an approval, final approval, or conditional approval, with or without modification of plats, including pre-liminary or final plats (as such terms are generally used or defined in section 276 of the Town Law, section 7-728 of the Village Law, and section 32 of the General City Law);

(6) in the case of an action involving a site plan, special permit, special use, con-ditional use, exception, variance, or similar special authorization, the action has received approval, with or without conditions or modifi-cations, by the appropriate local body (such as a legislative body, board of appeals, or plan-ning board), prior to September 1, 1977;

(7) in the case of a directly undertaken non-Type I local agency action, the chief fiscal officer of the local agency has certified on or before October 1, 1978 that substantial time, work or money were expended on the action prior to September 1, 1978:

(8) in the case of a state agency funded non-Type I action, the chief fiscal officer of each such state agency which is funding the action has certified on or before October 1, 1978 that the application for funding was ap-proved or in an approvable form prior to Sep-tember 1, 1978;

(9) in the case of a local agency funded non-Type I action, the chief fiscal officer of each such local agency which is funding the action has certified on or before October 1, 1978 that the application for funding was ap-proved or in an approvable form prior to Sep-tember 1, 1978;

166

(10) in the case of a non-Type I action involving a subdivision of land, the action has received prior to September 1, 1978 an approval, final approval, or conditional approval, with or without modifications of plats, including preliminary or final plats (as such terms are generally used or defined in section 276 of the Town Law, section 7-728 of the Village Law, and section 32 of the General City law);

(11) in the case of a non-Type I action involving a site plan, special permit, special use, conditional use, exception, variance, or similar special authorization, the action has received approval, with or without conditions or modifications, by the appropriate local body (such as a legislative body, board of appeals, or planning board) prior to September 1, 1978.

(d) Where an action is excluded pursuant to this section, but it is still practicable either to modify the action in such a way as to mitigate potentially adverse environmental effects or to choose a feasible and less environmentally damaging alternative, the commissioner may, at the request of any person or on his own motion, in a particular case or generally in one or more classes of cases, require the preparation of an EIS on such action pursuant to this Part.

The regulations also addressed the perennial problem of *lead agency designation*--so crucial to the quality of review given to projects involving more than one agency. Three criteria, in order of importance, were established for this purpose:

(1) whether the anticipated impacts of the action being considered are primarily of statewide, regional, or local significance, i.e., if such impacts are primarily of local concern, the local agency should be the lead agency;

(2) which agency has the greatest responsibility for supervising or approving the action as a whole, i.e., which agency has the more general governmental powers;

(3) which agency has the greatest capability for providing the most thorough environmental assessment of the action.

Provisions for two "consultation sessions" were also made in the new regulations for actions involving a private applicant. The first, following on the heels of California, insured the confidentiality of those elements of a project considered to be "trade secrets":

Section 617.16 Confidentiality.
A potential applicant to one or more agencies may, prior to applying for a permit, request a pre-application conference with all agencies that may be involved in determining lead agency. At such conference the potential applicant may identify those elements of the project which are in the nature of trade secrets or information, the nature of which if disclosed to the public or otherwise widely disseminated, would cause substantial injury to the competitive position of the potential applicant's enterprise. A potential applicant may request that such elements be kept confidential in accordance with the provisions of the Freedom of Information Law and other applicable laws.

The second provided for pre-EIS "scoping" sessions in a manner reminiscent of Massachusetts:

(h) Lead agencies shall make every practicable effort to involve applicants, other agencies and the public in the SEQR process. Consultations initiated by the lead agency can serve to narrow issues of significance and to identify areas of controversy, thereby focusing the subject matter of determinations and the scope and content of EIS's.

In June 1978, the Legislature amended SEQR for the third time in as many years. The amendments reflect the continuing concern over the administrative burden SEQR would place on local government:

Section 4. For the purposes only of simplifying procedures and clarifying the identification of actions and classes of actions that are likely to require preparation of an environmental impact statement as such matters are included in rules and regulations adopted pursuant to subdivision one of section 8-0113 of the environmental conservation law, the commissioner shall, wholly consistent with the provisions of article eight of the environmental conservation law, review such rules and regulations and adopt amendments to such rules and

regulations not later than the first day of
September, nineteen hundred seventy-eight.

Thus, while the amendments do not reduce the scope
of coverage, they do mandate the simplification of
procedures particularly with respect to non-Type I
actions.[117]
In response to this mandate, DEC proposed amendments to the rules and regulations intended to:

1. Simplify the procedures with regard to
   lead agency determination of non-Type I
   actions; and
2. Clarify the Type I list.

Both tasks are intended to make SEQR more usable at
the local level. In order that local government be
afforded an opportunity to familiarize itself with
the procedural changes, full implementation was again
postponed from September 1 to November 1, 1978.
Thus, although the amendments and the new regulations reflect an unwillingness to allow SEQR to at-
tain its full legislated applicability, they also
signal a trend (prevalent among other states) toward
concise, coordinated, and streamlined environmental
review.

5

Brief:
The Twelve Other
SEPA States

This Chapter contains detailed information about the environmental impact statement programs in the remaining twelve states not covered in Chapter 4. Information in this Chapter was derived from questionnaire responses verified, balanced, or supplemented where appropriate, by official documents, telephone contact, or correspondence.

The information on each state is divided into two main sections:

1. A statement of program evolution and current structure; and
2. A summary of currently available information concerning EIS program cost and caseload.

As in Chapter 4, detailed citations for legislation or regulation cited in the text have been omitted and may be found in the appropriate SEPA "Fact Sheet" in Appendix A.

CONNECTICUT

In 1972, Governor Meskill vetoed a bill for a state environmental policy act that would have required EIS's for private projects licensed by the state as well as projects carried out by state agencies. Later in the year, the Governor issued an executive order intended to achieve the "essential purposes of the legislation at less administrative expense."[118] Guidelines for the order were never formally adopted, however, and it consequently remained inactive.

To remedy this situation, the General Assembly enacted the Connecticut Environmental Policy Act based upon the language of the Governor's Executive Order. Signed in 1973, the Act was not to become effective until February 1, 1975 so that the Department of Environmental Protection (DEP) could have a chance to prepare the required guidelines for its implementation. Guidelines in the form of draft regulations were circulated in November of 1974, but objections to them were considered serious enough to warrant the postponement of further action until some administrative experience had been gained. In 1975, DEP requested an amendment postponing the effective date. The legislature balked at this and instead amended the Act to eliminate any reference to the preparation of regulations.[119] Efforts to resolve resultant procedural ambiguities through another amendment in 1976 were not successful and implementation of the Act was forced to commence with no specific requirements other than those expressed in the legislation.

Unlike the original SEPA proposal vetoed by the Governor, CEPA required EIS's only for state-initiated or state-supported projects that may "significantly affect the environment." Actions which may be "significant" were further defined as "includinq" those projects directly undertaken by state departments, institutions, or agencies, or funded in whole or in part by the state, which could have a "major impact" on the state's land, water, air or other environmental resources, or could serve short term to

172

the disadvantage of long-term environmental goals.[120]
Section 22a-16 of CEPA expanded upon the five
traditional EIS requirements of NEPA:

> (b) Each state department, institution or agen-
> cy responsible for the primary recommendation
> or initiation of actions which may significantly
> affect the environment shall in the case of each
> such action make a written evaluation of: (1)
> the consequences of the proposed action to the
> environment, including primary and secondary
> impacts on ecological systems; (2) adverse en-
> vironmental effects which cannot be avoided and
> irreversible and irretrievable commitments of
> resources should the proposal be implemented;
> (3) alternatives to the proposed action.
> (c) Each evaluation required by subsection (b)
> shall include an analysis relating the costs
> and benefits of the proposal over the short term
> to the costs and benefits over the long term.
> (Emphasis added)

The consideration of "primary and secondary impacts"
and the "costs and benefits of the proposal" were
significant additions to the language of NEPA. The
use of the term "written evaluation" in describing
the EIS, however, was of considerably less impact
than the "detailed statement" required by NEPA.
Separate legislation created a Council on Environmental Quality similar to the Federal CEQ.
While no provision for staff was made, the Council
was authorized under CEPA to review and make recommendations on EIS's. The Department of Environmental Protection, therefore, assumed informal responsibility for the implementation of the Act. Since no
Clearinghouse was established, DEP was to receive
copies of all EIS's for review and filing; final
recommendations concerning project review under CEPA
were made by the State Planning Council to the Governor, who made the final decision as to whether to
proceed with the project.
*Amendments to the law in 1977 significantly al-
tered the informal patterns of implementation which
had characterized CEPA since its initial passage.*
The amendments reassigned DEP the responsibility for
promulgating implementing procedures--this time in
the form of regulations:[121]

> Sec. 5. (NEW) Within six months of the ef-
> fective date of this act, the commissioner of
> environmental protection shall adopt regulations

173

to implement the provisions of sections 22a-1a
to 22a-1f, inclusive, of the general statutes,
as amended by this act. Such regulations shall
include: (1) specific criteria for determining
whether or not a proposed action may significantly affect the environment; (2) provision
for enumerating actions or classes of actions
which are subject to the requirements of this
act; (3) guidelines for the preparation of environmental impact evaluations, including the
content, scope and form of the valuations and
the environmental, social and economic factors
to be considered in such evaluations, and (4)
procedures for timely and thorough state agency and public review and for such other matters
as may be needed to assure effective public
participation and efficient implementation of
this act.

Despite the mandate for the preparation of detailed
implementing regulations, however, the amendments
chose to legislatively exempt both emergency measures and activities of a ministerial nature:

For the purposes of Section 22a-1b, actions
shall include but not be limited to new projects and programs of state agencies and new
projects supported by state contracts and
grants, but shall not include (1) emergency
measures undertaken in response to an immedi-
ate threat to public health or safety; or (2)
activities in which state agency participation
is ministerial in nature, involving no exercise
of discretion on the part of the state department, institution or agency.

The amendments also redefined the EIS requirement as a detailed written evaluation of environmental impact and dramatically refined its areas of
consideration:

(b) Each state department, institution or agency responsible for the primary recommendation
or initiation of actions which may significantly affect the environment shall in the case of
each such PROPOSED action make a DETAILED written evaluation of ITS ENVIRONMENTAL IMPACT BE-
FORE DECIDING WHETHER TO UNDERTAKE OR APPROVE
SUCH ACTION. ALL SUCH ENVIRONMENTAL IMPACT
EVALUATIONS SHALL BE DETAILED STATEMENTS SETTING
FORTH THE FOLLOWING: (1) A DESCRIPTION OF THE

174

PROPOSED ACTION: [(1)] (2) [The] THE ENVIRON-
MENTAL consequences of the proposed action [to
the environment], including [primary] DIRECT
and [secondary impacts on ecological systems]
INDIRECT EFFECTS WHICH MIGHT RESULT DURING AND
SUBSEQUENT TO THE PROPOSED ACTION:[(2)] (3) ANY
adverse environmental effects which cannot be
avoided and irreversible and irretrievable
commitments of resources should the proposal be
implemented; [(3)] (4) alternatives to the pro-
posed action, INCLUDING THE ALTERNATIVE OF NOT
PROCEEDING WITH THE PROPOSED ACTION: (5) MITI-
GATION MEASURES PROPOSED TO MINIMIZE ENVIRON-
MENTAL IMPACTS: (6), AN ANALYSIS OF THE SHORT
TERM AND LONG TERM ECONOMIC, SOCIAL AND ENVIRON-
MENTAL COSTS AND BENEFITS OF THE PROPOSED AC-
TION AND (7) THE EFFECT OF THE PROPOSED ACTION
ON THE USE AND CONSERVATION OF ENERGY RESOURCES.

The power of final review determination was reassigned to the Office of Policy and Management, an appendage to the Governor's Office. As such, OP&M may require the revision of any evaluation found to be inadequate.
    Perhaps the greatest significance was the ad-
dition of a unique *public hearing requirement* to the
already existing procedures for local notification:

THE DEPARTMENT, INSTITUTION OR AGENCY RESPON-
SIBLE FOR PREPARING AN EVALUATION SHALL PUBLISH
FORTHWITH A [Notice] NOTICE of [receipt and]
THE availability of such [evaluations and sum-
maries shall be published forthwith by such
town clerk] EVALUATION AND SUMMARY in a news-
paper of general circulation in the municipali-
ty at least once a week for three consecutive
weeks AND IN THE CONNECTICUT LAW JOURNAL. THE
DEPARTMENT, INSTITUTION, OR AGENCY PREPARING AN
EVALUATION REQUIRED BY SECTION 22a-1b, AS AMEND-
ED BY THIS ACT, SHALL HOLD A PUBLIC HEARING ON
THE EVALUATION IF TWENTY-FIVE PERSONS OR AN AS-
SOCIATION HAVING NOT LESS THAN TWENTY-FIVE PER-
SONS REQUEST SUCH A HEARING WITHIN TEN DAYS OF
THE PUBLICATION OF THE NOTICE IN THE CONNECTI-
CUT LAW JOURNAL.

COST AND CASELOAD

    Since CEPA has yet to be formally implemented
according to the new regulations (currently under
review), information on program costs reflects only

the stages of informal implementation between 1975 and 1977.

An interdepartmental message dated July 11, 1975 made several (probably somewhat inflated) estimates of initial CEPA-related costs within the Department of Environmental Protection (DEP). The Act requires DEP not only to write EIS's for projects they initiate or fund, but also to review the impact statements of all state-funded projects. The Department estimated two to four projects from the Department of Commerce, six to nine from the Department of Public Works, and six to eight from the Department of Transportation for a total of fourteen to twenty-one additional project reviews each year under CEPA from outside the Agency. A considerable number of projects within DEP were also anticipated as requiring impact evaluations:

PROJECT	COST
Parks & Recreation - 41	$335,000-preparation of EIS's
Fish and Waterlife - no number given but projects would include dam repairs, boat launching facilities, construction of fishways, removal of dams, lake and pond reclamation, stream improvement, control of aquatic vegetation, etc.	$282,000- lost federal funds due to delays
Forestry - 40-50 timber sales/year.	$100,000-lost in timber revenues from delays.
Wildlife - no number given but projects would include land clearing, thinning, construction of impoundments, parking lots, access roads, food plots, etc.	$350,000-lost federal funds due to inability to buy land and carry out projects.

Without at least two additional positions assigned to each departmental unit to carry out the terms of the Act, it was concluded, many of the core line programs would suffer:

Additional Manpower Necessary to Carry Out CEPA
According to Unit Directors

Water Resources	2 Water Resources Engineers 2 Sr. Environ. Analysts 2 Principal Environ. Analysts
Parks & Recreation	Use consultants to prepare EIS's review requirements uncertain but probably small.
Fish & Waterlife	2 Inland Fisheries Biologists 2 Marine Fisheries Biologists
Forestry	4 Professional Foresters
Land Acquisition	2 Senior Land Agents
Wildlife	2 Wildlife Biologists

No estimates from Air, Water Compliance, Solid Waste Management, Noise, Radiation, or Pesticides.

Thus, the estimate of initial CEPA implementation was upwards of $1 million for the first year. Since the implementation of the law was greatly hampered by the amendments in 1975, however, the actual cost of the program has been much less than the above estimates. In fact, only six projects have thus far required the preparation of an EIS under CEPA and four of these have been initiated by DEP. Of these four EIS's, two were contracted out at costs of $10,900 and $65,000, one was prepared in-house, and one is under contract negotiation. The two remaining EIS projects have both involved state-aid highway construction.[122]

The costs of CEPA coordination, like the caseload, have been small. Although it is estimated that these costs will increase substantially when the regulations become effective, DEP has invested only between six and twelve person-months per year in program coordination since 1975. Assuming a $15,000 per year salary and 140 percent overhead (includes clerical assistance),[123] *this amounts to an approximate yearly cost of between $18,000 and $36,000.*

HAWAII

Hawaii's EIS requirement began as an executive
order signed by the Governor in August 1971, and was
strictly limited to state projects. The review process was coordinated by the Office of Environmental
Quality Control (OEQC) in the Governor's Office,
which issued guidelines in October 1971. Several
bills were introduced over the next two years to
give the EIS requirement a statutory base.[124]

*In 1974, two separate statutes were passed which
together constituted a "state environmental policy
act."* Chapter 344 established a comprehensive environmental policy which included a unique entry concerning the *setting of population limits:*

[§344-3] Environmental policy. It shall be the
policy of the State, through its programs, authorities, and resources to:
(1) Conserve the natural resources, so that
land, water, mineral, visual, air and other
natural resources are protected by controlling
pollution, by preserving or augmenting natural
resources, and by safeguarding the State's
unique natural environmental characteristics in
a manner which will foster and promote the
general welfare, create and maintain conditions
under which man and nature can exist in productive harmony, and fulfill the social, economic, and other requirements of the people of
Hawaii.
(2) Enhance the quality of life by:
(A) Setting population limits so that the
interaction between the natural and man-made
environments and the population is mutually
beneficial;
(B) Creating opportunities for the residents
of Hawaii to improve their quality of life
through diverse economic activities which
are stable and in balance with the physical
and social environments;
(C) Establishing communities which provide
a sense of identity, wise use of land, ef-

ficient transportation, and aesthetic and
social satisfaction in harmony with the
natural environment which is uniquely
Hawaiian; and
(D) Establishing a commitment on the part
of each person to protect and enhance
Hawaii's environment and reduce the drain
on non-renewable resources.
(Emphasis added)

Chapter 343 established the environmental impact
statement requirement as follows:

[§343-4] Applicability and requirements. (a)
Except as otherwise provided, an environmental
impact statement shall be required for:
(1) Any action which will probably have significant effects and which proposes the use of
state or county lands or the use of state or
county funds, other than funds to be used for
feasibility or planning studies for possible
future programs or projects which the agency
has not approved, adopted, or funded, provided
that the agency shall consider environmental
factors and available alternatives in its
feasibility or planning studies.
(2) Any action within the classes of action
specified below:
    (A) All actions proposing any use within
    any land classified as conservation dis-
    trict by the state land use commission
    under chapter 205 which will probably have
    significant environmental effects.
    (B) All actions proposing any use within
    the shoreline area as defined in section
    205.31, or within 300 feet seaward of it
    which will probably have significant en-
    vironmental effects.
    (C) All actions proposing any use within
    any historic site as designated in the
    National Register or Hawaii Register as
    provided for in the Historic Preservation
    Act of 1966, Public Law 89-665, or chapter
    6, which will probably have significant
    environmental effects.
    (D) All actions proposing any use within
    the Waikiki-Diamond Head area of Oahu, the
    boundaries of which are delineated on the
    development plan for the Kalia, Waikiki,
    and Diamond Head areas (map designated as
    portion of 1967 city and county of Honolulu

179

General Plan Development Plan Waikiki-
Diamond Head (Section A)), which will
probably have significant environmental
effects.
(E) All actions proposing any amendments
to existing county general plans where such
amendment would result in designations other
than agriculture, conservation, or preservation, and which will probably have significant environmental effects, except all actions proposing any new county general plan
or amendments to any existing county general plan or amendments to any existing
county general plan initiated by a county.
(Emphasis added)

While EIS content was not specifically delineated in the legislation, the review mechanism was clearly defined (see Figure 5.1) and the mandate for rules
and regulations was explicit. An Environmental Quality Commission (EQC) was established in the Governor's Office and charged with the duties of developing rules and regulations and administering the Act,
much like the Council on Environmental Quality at the
Federal level. Members of the Commission serve without compensation and must include representatives
from labor, management, construction, environmental
interest groups, real estate groups, and architectural, engineering and planning representatives.
EQC is supported by a staff which is provided
by the Office of Environmental Quality Control; it
is the OEQC which reviews and comments on all EIS's
in the state while the Commission makes recommendations as to the acceptability of the statements prepared by state agencies. The final veto authority,
however, rests with the applicable chief elected
official:

(1) The governor, or his authorized representative, whenever an action proposes the use
of state lands or the use of state funds;
or
(2) The mayor, or his authorized representative,
of the respective county whenever an action
proposes only the use of county lands or
county funds.

The legislation also incorporated several innovative approaches into the EIS process. One such approach was a clause promoting the *incorporation of
previous EIS determinations by reference:*

FIGURE 5.1
GENERALIZED EIS PROCESS
HAWAII

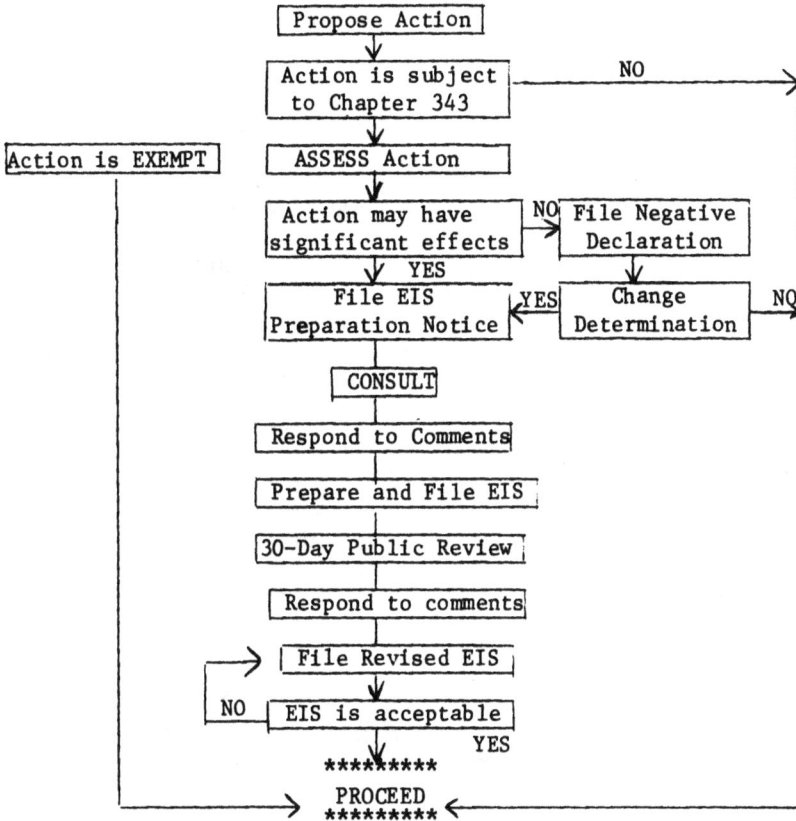

```
                    ┌─────────────────┐
                    │ Propose Action  │
                    └─────────────────┘
                             │
                             ▼
                    ┌─────────────────┐        NO
                    │ Action is subject├──────────────────────────►
                    │ to Chapter 343  │
                    └─────────────────┘
                             │
                             ▼
┌──────────────────┐ ┌─────────────────┐
│Action is EXEMPT  │ │ ASSESS Action   │
└──────────────────┘ └─────────────────┘
                             │
                             ▼
                    ┌─────────────────┐ NO ┌─────────────────┐
                    │ Action may have ├───►│ File Negative   │
                    │significant effects│  │ Declaration     │
                    └─────────────────┘    └─────────────────┘
                             │ YES                │
                             ▼                    ▼
                    ┌─────────────────┐ YES┌─────────────────┐ NO
                    │ File EIS        │◄───│ Change          ├──────►
                    │Preparation Notice│   │ Determination   │
                    └─────────────────┘    └─────────────────┘
                             │
                             ▼
                      ┌───────────┐
                      │ CONSULT   │
                      └───────────┘
                             │
                             ▼
                    ┌─────────────────┐
                    │Respond to Comments│
                    └─────────────────┘
                             │
                             ▼
                    ┌─────────────────┐
                    │Prepare and File EIS│
                    └─────────────────┘
                             │
                             ▼
                    ┌─────────────────┐
                    │30-Day Public Review│
                    └─────────────────┘
                             │
                             ▼
                    ┌─────────────────┐
                    │Respond to comments│
                    └─────────────────┘
                             │
                  ───►┌─────────────────┐
                      │ File Revised EIS│
                      └─────────────────┘
                             │
                  NO         ▼
              ◄────────┌─────────────────┐
                       │ EIS is acceptable│
                       └─────────────────┘
                             YES
                       *********
                        PROCEED
                       *********
```

(c) Whenever an agency proposes to implement an action or receives a request for approval, the agency may consider and, where applicable and appropriate, incorporate by reference in whole or in part previous determinations of whether a statement is required and previously accepted statements. The commission shall, by rules and regulations, establish criteria and procedures for the use of previous determinations and statements.

Another approach was the establishment of a statute of limitations on various types of judicial proceedings which could result from the EIS process.

[§343-6] Limitation of actions. (a) Any judicial proceedings, the subject of which is the lack of determination that a statement is or is not required for a proposed action not otherwise exempted, shall be initiated within 180 days of the agency's decision to carry out or approve the action, or if a proposed action is undertaken without a formal determination by the agency that a statement is or is not required, a judicial proceeding shall be instituted within 180 days after the proposed action is started.

(b) Any judicial proceeding, the subject of which is the determination that a statement is or is not required for a proposed action, shall be initiated within sixty days after the public has been informed of such determination pursuant to section 343-2.

(c) Any judicial proceeding, the subject of which is the acceptability of a statement, shall be initiated within sixty days after the public has been informed pursuant to section 343-2 of the acceptance of such statement; provided that only affected agencies, or persons who will be aggrieved by a proposed action and who provided written comments to such statement during the designated review period shall have standing to file suit; further provided that contestable issues shall be limited to issues identified and discussed by the plaintiff in the written comments.

After consultation with the affected agencies and four months of public hearings, rules and regulations were adopted in 1975; this was important since the legislation deferred to the regulations on most of the substantive issues. The regulations outlined specific procedures for public review and for agency response to comments. Of particular interest is the special care taken to assure proper availability of EIS's to interested parties:

1:51 DISTRIBUTION. The Commission shall be responsible for the publication of the notice of availability of the EIS in its bulletin, and for distribution of the EIS for agency and public review. The Commission shall develop a list of reviewers (i.e., persons and agencies with jurisdiction or expertise in certain areas relevant to various actions) and a list of public depositories where copies of

EIS's will be available and to the extent
possible, the Commission shall make copies
of the EIS available to individuals requesting same by telephone, in writing, or in
person. To minimize postage, the Commission may make special arrangements for direct distribution of EIS's by the approving
agency to particular recipients.

Thus, Hawaii's program established a highly visible
and easily entered process of environmental review.
By vesting the final approval powers with the Governor, it also separated the advocacy of projects
from the ultimate evaluation of environmental impacts.

COST AND CASELOAD

The cost of coordinating Hawaii's environmental
impact statement program has been fairly consistent
since the implementation of its rules and regulations in 1975. In fiscal 1975-76, the cost of administering the program and reviewing State EIS's
was about $80,000; this was broken down as fol-
lows:[125]

Administrative Costs ($35,000)	Review Costs ($45,000)
2 professional staff	2 professional staff
¼ clerical staff	¼ clerical staff

Included under the administrative costs were $10,000
for hearings on the rules and regulations and $3,500
for handbooks and training seminars.

In fiscal 1976-77, the planned cost of this
function rose to $120,000. About 2½ person-years
would be allocated to the Environmental Quality Commission in the form of staff support. Another 2½
person-years would be allotted within the Office of
Environmental Quality Control for review and analysis
of EIS's. Supervision, administrative and support
services would require 3 person-years. Concomitant
costs for printing, postage, supplies, travel, etc.,
would also be incurred. One new feature of support
for the Commission was the provision of $15,000 for
the services of a deputy attorney general.[126] The
legal services, however, have since been terminated.

In fiscal 1976-77, twenty-six EIS's were prepared for state projects and seven EIS's for private
projects in compliance with Hawaii's impact statement law. At least ten times this number of Negative

Declarations were also filed during the same period.[127] The only other information pertaining to
caseload comes from the State Department of Transportation under which fifteen EIS's were prepared
in fiscal 1975-76. No figures on EIS preparation
costs were available since they were not internally
distinguished from total project costs.

INDIANA

The Indiana Environmental Policy Act, approved in February 1972, was a near verbatim copy of NEPA. It contained an identical definition of environmental impact statement content and also included NEPA's "Environmental Bill of Rights," with the substitution of "Indiana State government" for "Federal government" in the wording. Unlike NEPA, however, Indiana's law specifically limited the applicability of the EIS requirement to actions directly undertaken or supported by State agencies:

> Sec. 6. Nothing in this chapter shall be construed to require an environmental impact statement for the issuance of a license or permit by an agency of the state.

An interagency body--the Environmental Management Board (EMB)--is responsible for coordinating the impact statement program in addition to a wide range of other environmental responsibilities. EMB is composed of eleven members including representatives of state agencies, labor, industry, and the public at large. Although the threshold determination (whether or not an EIS is required) is made by the agency proposing the action, EMB has prescribed the criteria for making this decision through regulations issued in 1975. Furthermore, all EIS's prepared must be submitted to EMB for its acceptance, acceptance with recommendations, or rejection according to regulations for the EIS review process issued by EMB in January 1977. The criteria for this decision, however, is largely *procedural* (conformance with the regulations); since EMB conducts no substantive review of its own, it depends on other agencies, organizations, and individuals to comment on specific issues. *Thus, Indiana's program departs from the NEPA model by vesting the ultimate approval power in a third party, but fails to extend this process past the purely procedural matters of compliance.*

Although the law was passed in 1972, it remained essentially unimplemented until early 1975 when EMB issued regulations specifying criteria for threshold determination. Thus, up to this point, the cost of the impact statement program was negligible since the responsibilities were incremental to an already existing board of unpaid representatives (EMB).

In fiscal 1976, a senior staff member was assigned to the Technical Secretary of the Agency to assist in the coordination and execution of SEPA responsibilities. The cost of this basically administrative function was $27,600. However, only a fraction of the staff member's time is devoted to these responsibilities. EMB also requested the addition of a full-time position for the review of EIS's, but this position was never provided.[128]

Since the main procedural regulations for the EIS review process only became effective in January 1977, the program is still in its infancy. Although a dozen impact statements have been prepared since the passage of the Act, all of these have entailed compliance with NEPA as well as the Indiana EPA (primarily highway projects). *Thus, there have been no EIS's prepared exclusively in compliance with the state law.*[129]

MARYLAND

The Maryland Environmental Policy Act (MEPA), passed in 1973, constitutes a significant departure from the NEPA model. *MEPA covers only "proposed state actions" and its major purpose is to provide the State legislature with information about environmental effects consequent to appropriations of State funds and legislative proposals.* Thus, it is rather limited in its applicability in comparison to NEPA. Nonetheless, several alterations in the traditional EIS requirement make Maryland's law worthy of mention.

The initial legislation required an "environmental effects report" (EER) on each proposed state action significantly affecting the environment, *natural as well as socio-economic.* Amendments to the law in 1975 extended this to include *historic* environments as well. A "proposed state action" was defined very narrowly as "requests for legislative appropriations or other legislative actions that will alter the quality of the air, land, or water resources. It does not include a request for an appropriation or other action with respect to the rehabilitation or maintenance of existing secondary roads." Furthermore, the content of EER's was defined in a manner greatly differing from the traditional impact statement:

> EER's are to include, but not be limited to, discussions of:(1) the effects of the proposed action on the environment, including adverse and beneficial environmental effects that are reasonably likely if the proposal is implemented or if it is not implemented; (2) Measures that might be taken to minimize potential adverse environmental effects and maximize potential beneficial environmental effects, including monitoring, maintenance, replacement, operation, and other follow-up activities; and (3) Reasonable alternatives to the proposed action that might have less adverse environmental effects or greater beneficial environmental

effects, including the alternative of no
action. (Emphasis added)

The legislation designated the Department of
Natural Resources (DNR) as responsible for the preparation of implementing guidelines but assigned it
no power over the review process. The guidelines
issued by DNR in December of 1973 and revised in
June of 1974 provided departments of the state with
a preliminary environmental assessment form (EAF)
with which to determine the significance of a proposed action. *If the impact is determined to be
significant, either adverse or beneficial, an EER
is required.* The preparation, circulation, and review of the document then follows procedures established by the individual agencies, so long as these
procedures are consistent with the general DNR guidelines. The body of every EER, for instance, must
contain the following sections:

1. Summary
2. A description of the proposed action
3. Environmental (ecologic and socio-economic)
   setting without the action
4. Adverse and beneficial environmental effects
5. Measures taken to minimize adverse environmental effects and maximize beneficial environmental effects
6. Any adverse environmental effects which
   cannot be avoided should the action be implemented
7. Alternatives to the proposed action
8. Coordination with other interested parties

The guidelines also mandated that all EER's
and EAF's be available for public inspection or pur-
chase at cost of reproduction. The State Clearinghouse, Department of State Planning (DSP) is responsible for the distribution of environmental documents. After screening for expertise, documents are
distributed to selected state departments as well as
to affected local jurisdictions and other interested
parties. DSP also prepares a quarterly list of all
such documents and distributes it to:

1. All members of the State General Assembly;
2. All State departments;
3. All newspapers, private citizens and citizen groups who request such listing.

The guidelines also incorporate a unique provision
for *public meetings:*

D. Public Information Meetings: A public information meeting on Environmental Assessment Forms and Environmental Effects Reports may be held by the action unit upon
formal written request from a local governmental unit or from 50 or more citizens.
(Emphasis added)

Before collecting and returning review comments,
DSP conducts its own review of EER's to determine if
sufficient information is presented to enable proper
review and to make recommendations to the proposing
agency based on their findings. The ultimate determination of adequacy, however, rests with the agency
initiating the action.

*Thus, although the Maryland law contains many
interesting improvements over customary language,
like NEPA, it lacks a really strong administrative
power to enforce the Act.*

COST AND CASELOAD

There has been no explicit attempt to assess
the costs of preparing impact statements under the
Maryland Environmental Policy Act despite the fact
that 156 preliminary assessments and 20 EER's were
filed by state agencies under MEPA in 1976.

The costs of coordinating the Act are assumed
jointly by the Department of Natural Resources (DNR)
and the Department of State Planning (DSP). Since
the preparation of revised guidelines in 1974, the
cost to DNR of administering the Act has been only
one to two person-days per month.[130] Similarly, the
costs to DSP of executing its Clearinghouse responsibilities under the Act have been negligible; it
expends only $7,000-$10,000 per year (¼ professional
person and ¼ clerical person) in the distribution,
collection, and review of environmental documents
under MEPA.[131]

No figures were available concerning the cost
of document preparation and review.

189

MICHIGAN

A directive to department heads issued by the
Governor in December 1971, concerning environmental
review was superseded by formal executive orders in
1973 and 1974. The latter, Executive Order 1974-4,
issued by Governor Millikan in May 1974, required
state agencies to prepare environmental impact
statements on proposed major actions within their
jurisdictions "that may have a significant impact
on the environment or human life." *Thus, Michigan's
approach augments that of NEPA in that considerations
of human ecology are clearly articulated.* The human
ecological approach takes into account the reality
of human dominance in ecosystems that can be politi-
cally and geographically defined. Emphasis is placed
on the organization and technology which is employed
by humans to control the flow of resources and wastes
in an environment of limited capacity.[132] This emphasis is further evidenced by the contents prescribed for EIS's prepared under the Order:

> Each statement shall contain the following:
> (1) A description of the probable impact of
>   the action on the environment, including
>   any associated impacts on human life.
> (2) A description of the probable adverse ef-
>   fects of the action which cannot be avoid-
>   ed (such as air or water pollution, threats
>   to human health or other adverse effects on
>   human life).
> (3) Evaluation of alternatives to the proposed
>   action that might avoid some or all of the
>   adverse effects, including an explanation
>   why the agency determined to pursue the ac-
>   tion in its contemplated form rather than
>   an alternative.
> (4) The possible modifications to the project
>   which would eliminate or minimize adverse
>   effects, including a discussion of the ad-
>   ditional costs involved in such modifica-
>   tions.

Prior to 1973, the EIS program was coordinated by the Department of Natural Resources. The Executive Order in 1973, however, shifted the responsibilities to a newly created "Environmental Review Board" whose stated function was to coordinate the state environmental impact review program including the determination of those actions of state agencies that should be suspended or modified, and to make requests for appropriate agency action in response to that determination. The Board consists of seventeen members appointed by the Governor. Ten members are appointed from the general public and the remaining seven members as follows:

- The Attorney General
- Director, Department of Agriculture
- Director, Department of Commerce
- Director, Department of Natural Resources
- Director, Department of Public Health
- Director, Department of State Highways
  and Transportation
- Director, Department of Management and Budget

The Executive Order in 1974, formalized a *two-step review process* by assigning technical review responsibilities to an Inter-Departmental Environmental Review Committee (Intercom). The Committee, which consists of one member appointed from each State Department, generally forwards its findings to the Board for final approval; Intercom may, however, return an EIS it deems inadequate for redrafting.

Guidelines issued pursuant to the Executive Order in November 1975 greatly clarify both the review process and the extensiveness of coverage. "Major state action," for example, was defined as "any policy, administrative action, or project;" the term "administrative action" was defined as including "the issuance of licenses, permits, or other forms of approval...land acquisition, disposition or leasing, or construction that will utilize state funds including grants-in-aid." Thus, the Guidelines interpret Executive Order 1974-4 as applying not only to actions undertaken directly by state agencies, but also to activities funded or approved by those agencies.

The review process (see Figure 5.2) outlined by the Guidelines provides a sixty day minimum public review period and allows Intercom forty days for the technical review of EIS's. Except as previously noted, Intercom's recommended course of action is forwarded to the Board by majority vote. Each EIS is

FIGURE 5.2
MICHIGAN'S EIS PROCESS

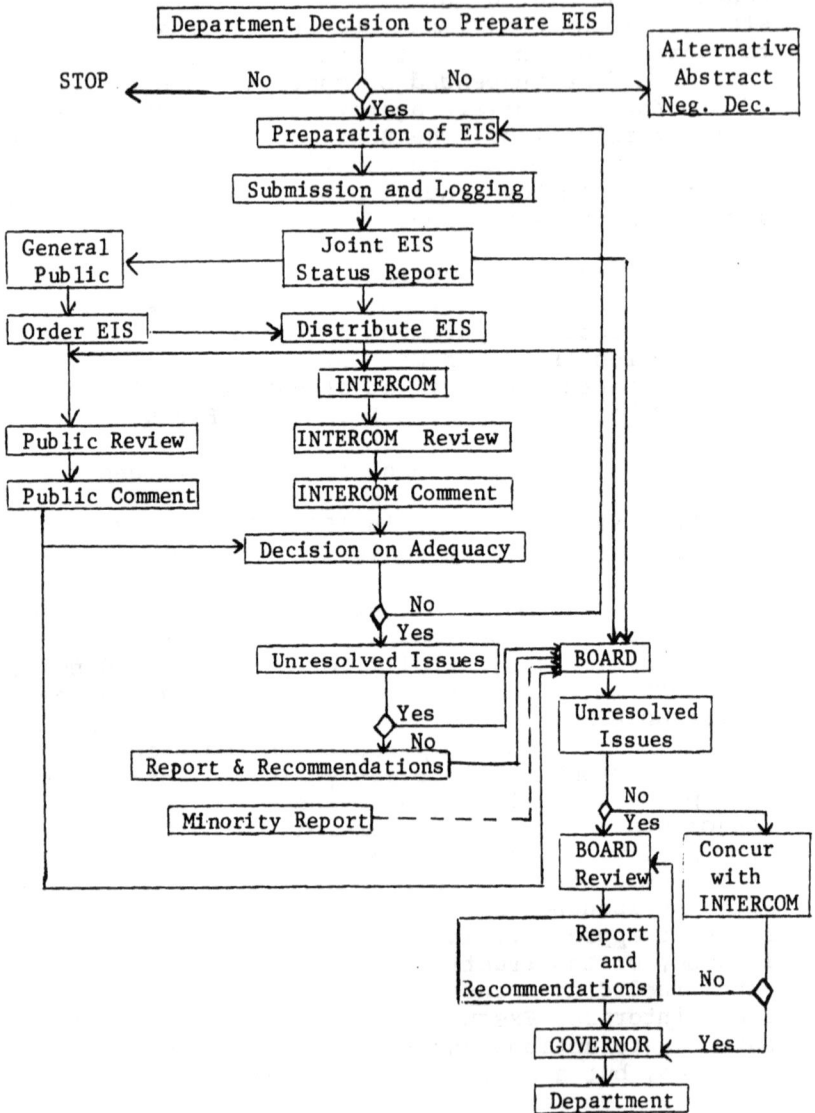

FIGURE 5.2
MICHIGAN'S EIS PROCESS

reviewed to determine whether it contains adequate discussion and analysis of the following set of potential impacts:

(1) A potential significant impact on the human environment that could adversely affect the public health and welfare or could degrade the quality of life.
(2) Alteration or destruction of a significant element of the human, natural, amenity or historic resources of the State.
(3) Significant alteration of existing land use patterns.
(4) Significant alteration of population distribution of which would lead to potential distribution changes.
(5) Significant impact on the maintenance and enhancement of the long-term productivity of the State's natural resources.
(6) The imposition of an alteration to the ecological balance of a significant element of the environment.
(7) Significant additional uses of energy resources or the acquisition thereof.

The Board then acts to either concur with or overrule Intercom's recommendations by a majority vote. The Board makes any recommendations on the appropriateness of proposed actions to the Governor. In addition, the Board may recommend to the Governor that an EIS be prepared on the basis of public concern or controversy.[133] Although it is, of course, questionable whether review bodies composed largely of agency representatives will be quick to reject impact statements of its members, the existence of the ability to return EIS's for redrafting is an important element of the Michigan Executive Order.

COST AND CASELOAD

Executive Order 1974-14 mandated that the Department of Management and Budget provide an Executive Secretary for the Environmental Review Board and additional staff support as may be needed. The Executive Secretary and a part-time secretary were initially allocated to the program. However, growing responsibilities required the addition of another professional in 1975. The budget for fiscal 1975-76 and 1976-77 held steady at $60,000; this figure did not include the nineteen agencies' time alloted to Intercom review. Table 5.1 depicts the proposed budget for review coordination and administration under Executive Order 1974-4 for fiscal 1977-78.

TABLE 5.1
MICHIGAN ENVIRONMENTAL REVIEW BOARD
INTERDEPARTMENTAL ENVIRONMENTAL REVIEW COMMITTEE
PROPOSED BUDGET FY 1977-78

Salaries, Wages & Fringes (2 professionals, 1 stenographer)	$61,738
Telephone	1,000
Subscriptions	20
Postage	2,000
Board Travel	5,000
Office Supplies, Printing, Materials, etc.	3,500
TOTAL	$73,258

The costs of preparing EIS's under the Executive Order have been minimal since most information contained in the documents were required, in one form or another, prior to the Order.[134] Indeed, a large percentage of the impact statements submitted to the Board are joint State-Federal documents which would be required under NEPA in the absence of the state review system. Whatever the incremental costs of document preparation are, however, they are decreasing, since fewer EIS's are being prepared pursuant to the state requirement; in calendar 1975, for example, thirty-five EIS's and thirty-four Negative Declarations were reviewed and acted upon by the Board. In 1976, these numbers dropped to twenty-two EIS's and ten Negative Declarations and in 1977, to fifteen EIS's and fifteen Negative Declarations.[135]
Although EIS costs in Michigan have not, as yet, been formally assessed, such costs should be documented in the near future in response to recent legislative mandate.[136]

MONTANA

The Montana Environmental Policy Act, effective
March of 1971, was another verbatim transposition of
the language of NEPA. The sections dealing with purpose, policy, and the environmental impact statement
requirement nearly parallel the contents of NEPA's
Title I. The Montana Act, however, is stronger than
NEPA in an important way; in its parallel section to
NEPA's Section 101(c) (which states that each person
"should enjoy a healthful environment"), MEPA states
that "the legislative assembly recognizes that each
person shall be entitled to a healthful environment."
(Emphasis added) Also, under Montana's Constitution,
citizens have a *right to a healthful environment*.

MEPA established an Environmental Quality Council (EQC) in the legislative branch, which was charged with developing state environmental policy goals,
promulgating EIS guidelines, and evaluating operating programs, proposals, and actions in the environmental field. A Montana Commission on Environmental
Quality (MCEQ) was also formed in the executive
branch with the primary responsibility of promulgating model EIS rules.

With the assistance of the EQC, MCEQ adopted
final rules for implementing MEPA in January 1976.
Individual agencies were required to implement the
law in accordance with these uniform rules. Probably the most significant feature of the MCEQ Rules
is the development of a procedure for "Preliminary
Environmental Reviews"(PER). This allows each agency to first determine whether or not a proposed project is a "major" action:

> (2) A PER shall be prepared by the Department
> on all proposed actions of the Department or
> Board, other than those described in subsection
> (1) of this rule or where the action is clearly
> a major state action having a significant im-
> pact on the human environment, thereby requir-
> ing the preparation of an EIS, on which a de-
> termination must be made as to the significance
> of its effect on the environment. If the PER

shows a potential significant effect on the human environment, an EIS shall be prepared on that action.

(3) The following are actions which normally require the preparation of an EIS:

    (a) the action may significantly affect environmental attributes recognized as being endangered, fragile, or in severely short supply;

    (b) the action may be either significantly growth inducing or inhibiting; or

    (c) the action may substantially alter environmental conditions in terms of quality or availability.

(4) The Department shall maintain a list of those activities or functions that fall within the categories described in the preceding subsections of this rule. The list shall be maintained as a public document and copies of the list and any subsequent revisions sent to the MCEQ, the EQC, and any member of the public who has requested a copy of the list. The MCEQ, the EQC, or any member of the public may recommend additions to or deletions from the list. The Department shall review the recommendations for additions to or deletions from the list and advise the person or group making the recommendation in writing of the reasons why the recommended additions or deletions were or were not made.

The rules also provide categorical exemptions for all "ministerial actions," "existing facilities," and "investigation and enforcement" actions.

    Since each agency of the state is responsible for developing its own procedures for implementing the law (consistent with the general guidelines), the reaction has been anything but uniform. In both the Department of State Lands and the Department of Health, for example, one person is responsible for coordinating the program and preparing EIS's on all projects which require them. In contrast, the Departments of Natural Resources and Conservation (DNRC) and Highways have a more integrated approach. In Highways, the bureau staff people responsible for technical review of the proposed action are also responsible for EIS preparation. Special environmental staffs are available to assist bureau people with nontechnical aspects and with interagency coordination.[137]

    Amendments to MEPA in 1975 explicitly extended the EIS requirement to the private sector by prescribing a system of applicant fees:

69-6518. Fee may be imposed. (1) Each agency
of state government charged with the responsibility of issuing a lease, permit, contract,
license or certificate under any provision of
state law may adopt rules prescribing fees
which shall be paid by a person, corporation,
partnership, firm, association, or other private entity when an application for a lease,
permit, contract, license, or certificate will
require an agency to compile an environmental
impact statement as prescribed by Section 69-
6504, R.C.M. 1947, of the Montana Environmental
Policy Act. An agency must determine within
thirty (30) days after a completed application
is filed whether it will be necessary to compile an environmental impact statement and assess a fee as prescribed by this section. The
fee assessed under this section shall only be
used to gather data and information necessary
to compile an environmental impact statement as
defined in the Montana Environmental Policy Act.
No fee may be assessed if an agency intends
only to file a negative declaration stating that
the proposed project will not have a insignificant impact on the human environment.
(2) In prescribing fees to be assessed against
applicants for a lease, permit, contract, license, or certificate, as specified in subsection (1), an agency may adopt a fee schedule
which may be adjusted depending upon the size
and complexity of the proposed project. No
fee may be assessed unless the application for
a lease, permit, contract, license, or certificate will result in the agency incurring expenses in excess of two thousand five hundred
dollars ($2,500) to compile an environmental
impact statement. The maximum fee that may be
imposed by an agency shall not exceed two per
cent (2%) of any estimated cost up to one million dollars ($1,000,000); plus one per cent
(1%) of any estimated cost over one million
dollars ($1,000,000) and up to twenty million
dollars ($20,000,000); plus one-half of one per
cent (½ of 1%) of any estimated cost over twenty million dollars ($20,000,000), and up to one
hundred million dollars ($100,000,000); plus
one-quarter of one per cent (¼ of 1%) of any
estimated cost over one hundred million dollars ($100,000,000) and up to three hundred
million dollars ($300,000,000); plus one-eighth
of one per cent (1/8 of 1%) of any estimated

cost in excess of three hundred million dollars ($300,000,000). If an application con-
sists of two (2) or more facilities, the filing
fee shall be based on the total estimated cost
of the combined facilities. The estimated
cost shall be determined by the agency and the
applicant at the time the application is filed.

Despite the legal extension of the law, however, no
equally strong attempt has been made to coordinate
the preparation of EIS's with the permit process.
Thus, the EIS may become perceived as more of a burden to private developers.

COST AND CASELOAD

An estimate of MEPA coordination costs is somewhat simplified by the fact that EQC was specifically
created for this purpose (among others). Staff within the Environmental Quality Council has grown as
follows:

	1971	1972	1973	1974	1975	1976	1977
				TABLE 5.2 EQC STAFFING			
Admin.	1	1	1	1	1	1	1
Research	2	2	4	4	4	4	4
Clerical	1	1	2	2	2	2	2
Total Salary	$27,136	$36,113	$69,070	$90,113	Not Avail.	$97,899	Not Avail.

In fiscal 1976, the EQC had a budget of $139,582.
For fiscal 1977, the total EQC budget appeared as
follows:

ENVIRONMENTAL QUALITY COUNCIL
OPERATIONAL PLAN/BUDGET
FISCAL YEAR 1977-78

Salaries	$88,461	
Other Compensation	2,000	
Employee Benefits	12,385	
Total Personal Services		$102,846
Contracted Services	12,385	
Supplies & Materials	1,300	
Communications	4,000	
Travel	10,000	
Rent	9,200	
Repair & Maintenance	350	
Other Expense	1,500	

```
Total Operating Expense                    $ 38,735
Equipment                    1,000
  Total Equipment                          $  1,000
Total Program Costs                        $142,581
```

It was estimated, however, that only 20 percent of
the EQC staff time was spent on MEPA-related matters.[138] Thus, a good estimate of MEPA coordination
costs for fiscal 1976 is (.20)($139,582.55) or about
$28,000. In 1977, this increased proportionally to
approximately $31,200.

In fiscal 1974, thirty-one EIS's were prepared
in compliance with MEPA at the state level. The
next three years, this number dropped to between ten
and twenty (see Table 5.3). It is interesting to
note that nearly 200 preliminary assessments resulted in only 10 EIS's in 1975 for a ration of 20 Negative Declarations for every EIS prepared.[139]

TABLE 5.3			
DOCUMENTS SUBMITTED IN COMPLIANCE WITH MEPA			
Agency	EIS/PER FY 1975	EIS/PER FY 1976	EIS/PER FY 1977
Department of Fish & Game	1/4	0/4	1/17
Department of Health and Environmental Sciences	1/49	3/33	3/44
Department of Highways	1/138	5/112	0/49
Department of State Lands	2/2	4/6	3/9
Department of Livestock	1/0	--	--
Department of Natural Resources and Conservation	4/0	7/2	1/17
TOTAL	10/193	19/157	9/119

The major project-originating agency, the Department of Natural Resources and Conservation (DNRC),
maintains a core team of three persons with back-
grounds in technical writing, resource economics, and
earth sciences for preparation and review of EIS's.
Expertise in other areas (hydrology, engineering,
etc.) is available to the core team in departmental
divisions at the request of the director.[140] Because of the numerous positions involved in varying
degrees of participation in the EIS process, it was
not possible to derive a concrete estimate of the
costs incurred in preparing, reviewing, and administering the EIS mandate. Agency staff hours expended in review of EIS's varied from less than one
person-hour to perhaps sixty to eighty. Likewise,
the time spent in preparing EIS's ranged from five

people working three weeks to twenty-five people
working two years. Assuming an average salary of
$13,500 per year, this means that EIS review costs
vary from $10 to $500 per document, and EIS preparation costs range from about $4,000 to $675,000, depending on the scope of the project.[141]

NEW JERSEY

In October 1973, Governor Cahill issued Executive Order #53. This order directed "all Departments
and Agencies of the State" to assess the environmental impact of their projects. *Projects included under*
this mandate were delineated specifically:

Any construction project with a total cost
greater than $1,000,000.
Any construction project with a total cost of
less than $1,000,000 which, by reason of its nature,
location in a fragile or undeveloped area, or method
of construction or operation, has the potential for
substantial adverse environmental impact.
Construction projects undertaken by local,
county or regional governments or agencies for which
a department or agency of the State has provided
funding in excess of $1,000,000.
Construction projects undertaken by local,
county or regional governments or agencies for which
a department or agency of the State has provided
funding of less than $1,000,000, but which, by reason
of the project's nature, location in fragile or undeveloped area, or method of construction or operation, has the potential for substantial adverse environmental impact.

The Department of Environmental Protection (DEP) also
requires, as a matter of policy, an EIS for private
actions requiring multiple permits from DEP. Once
environmental review has been completed, a private
project receives "conceptual" approval and is released to the line agencies within DEP responsible
for permit issuance. The affected areas of the private sector include heavy industry, utilities, and
major commercial or residential development.[142]
The responsibility for coordination and review
under the Executive Order was assigned to the Office
of Environmental Review (OER) within the Department
of Environmental Protection. OER was originally established by administrative order in response to the
need for Departmental coordination in environmental

review under NEPA. It also coordinates two independent EIS programs under the Wetlands and Coastal Area Facilities Review Acts.

The Office of Environmental Review is charged with the responsibility for:

(a) establishing and maintaining the Department policy for the environmental impact statement process;

(b) preparing environmental impact statement guidelines and submission procedures, internal review procedures;

(c) coordinating the review of all environmental assessments and impact statements;

(d) serving as staff for the State Historic Preservation Officer on cultural resource reviews.

In order to meet these various responsibilities, the Office has assembled an interdisciplinary team of environmental specialists including the disciplines of biology, engineering, history, archaeology, and land-use planning.[143]

Under EO #53, the project sponsor is required to prepare a brief environmental assessment for submission to DEP. The OER then reviews the assessment and within thirty days must make a determination as to the need for the sponsor to prepare a full EIS. If an EIS is deemed necessary, DEP (through its line agencies and OER) must conduct a full review within four weeks of the receipt of the statement and report its findings and recommendations to the State Planning Task Force. Within four weeks of DEP's review, the proposing agency must notify the State Planning Task Force in writing of its response to DEP's analysis; where recommendations of the Department are not accepted by the proposing agency, it must explain its action. The State Planning Task Force then must consider and reconcile the differences between DEP and the proposing agency. *A project cannot proceed until such reconciliation takes place.*

Thus, New Jersey has separated project initiation from project review in a most interesting way-- the power of threshold determination rests in the hands of a separate review entity (OER) while final review authority of EIS's is vested in a mediator (the State Planning Task Force) which reconciles differences between project sponsor and reviewer.

202

COST AND CASELOAD

During fiscal 1977, a total of sixty-three environmental documents were processed through the OER as compared to thirty-five in fiscal 1974. A breakdown of these documents, however, yields interesting results:

		TABLE 5.4	
		NEW JERSEY CASELOAD	
	NEPA	EO #53	DEP/EIS (private)
1974	19	5	11
1977	34	28	1

Thus, the total number of documents submitted for <u>government</u> projects has increased dramatically while the number of <u>private</u> assessments has dropped. *Furthermore, the number of full EIS's actually being prepared has declined both under NEPA and the state requirement.*[144] In 1977, for example, of the twenty-eight documents prepared under EO #53, twenty-six were Negative Declarations and only two were full EIS's.

The Office of Environmental Review maintains a staff of seven as follows:

- 1 Chief, Environmental Review
- 2 Environmental Review Specialists
- 3 Cultural Resource Specialists
- 1 Secretary

The Office operates on approximately $125,000 per year including salaries and overhead.[145] The three Cultural Resource Specialists (or about $50,000 of this total), however, are occupied exclusively by the National Historic Preservation Program and of the remaining $75,000, expenditures are split about 60 percent for NEPA review and 40 percent for EO #53. Thus, the cost of administering the state EIS program and reviewing documents within OER is approximately $30,000 per year.

Departmental line agencies conduct secondary review within the area of their technical expertise and regulatory jurisdiction. The cost of this additional review is estimated at $25,000 per year[146] bringing the total cost of administration and review to about $55,000 per year.

The cost of document preparation is highly variable given the diverse nature of project proposals. For example, an initial assessment ending in a

Negative Declaration generally costs between $300 and $500 to prepare while a single EIS has cost as much as $200,000. On the average, however, state EIS's range between $10,000 and $20,000 each to prepare.[147] Thus, for 1977, the cost of document preparation can be estimated:

$$26 \text{ Negative Declarations} \times \$450 = \$11,700$$
$$2 \text{ EIS's} \times \$15,000 = \$30,000$$
$$\text{Total} \quad \$41,700$$

Therefore, the total cost of the state EIS program under Executive Order #53 is approximated by summing the component costs:

$30,000/yr (OER admin. and review)
$25,000/yr (line agency review)
$41,700/yr (document preparation)
$96,700/yr TOTAL

SOUTH DAKOTA

The South Dakota Environmental Policy Act of 1974 was limited to the establishment of an impact statement requirement; *it made no statement of general environmental policy nor was the overall purpose of the Act clearly defined.* Furthermore, the applicability of the legislation with respect to permitted actions was obscured by two seemingly contradictory clauses. Section 11-1A-4 of the Act stated that:

> All agencies shall prepare, or cause to be prepared by contract, an environmental impact statement or any major action they <u>propose or approve</u> which may have a significant effect on the environment. (Emphasis added)

The section immediately preceding the above, however, <u>exempted</u> several classes of action including those actions requiring the approval of regulatory agencies:

> 11-1A-3. Actions not subject to chapter. As used in this chapter, unless the context otherwise required, "actions" do not include: (1) Enforcement proceedings or the exercise of prosecutorial discretion in determining whether or not to institute such proceedings; (2) Actions of a ministerial nature, involving no exercise of discretion; (3) Emergency actions responding to an immediate threat to public health or safety; (4) Proposals for legislation; or (5) <u>Actions of an environmentally protective regulatory nature.</u> (Emphasis added)

While the applicability of the EIS prerequisite was unclear, the law specifically delineated the required contents of impact statements:

> 11-1A-7. Contents of environmental impact statement. An environmental impact statement

shall include a detailed statement setting
forth the following:
(1) A description of the proposed action and
its environmental setting;
(2) The environmental impact of the proposed
action including short-term and long-term
effects;
(3) Any adverse environmental effects which
cannot be avoided should the proposal be implemented;
(4) Alternatives to the proposed action;
(5) Any irreversible and irretrievable commitments of resources which would be involved in
the proposed action should it be implemented;
(6) Mitigation measures proposed to minimize
the environmental impact; and
(7) The growth-inducing aspects of the proposed
action.

Since no single agency was assigned the legal
responsibility for administering the law, the Department of Environmental Protection (DEP) assumed
this role by issuing informal guidelines for the implementation of the Act in 1974. The major purpose
of these guidelines was to suggest criteria for determining what constitutes a "major action" or "significant environmental impact." This was accomplished through a suggested environmental assessment form
(EAF); the ultimate responsibility for establishing
implementing procedures, however, resided with the
individual state agencies.
Thus, the South Dakota EPA requires environmental impact statements on actions of state agencies
having significant environmental impacts and the
statements must cover the seven areas identified in
the Act. EIS's are to be prepared in draft form to
allow citizen input before being filed in final form
with the Department of Environmental Protection. The
Act does not provide authority for the Department to
specify criteria for implementation, to review statements for adequacy (other than meeting the specific
requirements of the Act), or to enter the agency decision-making process. The DEP functions strictly
as a review agency and a repository for draft and
final impact statements.[148]

COST AND CASELOAD

Since its passage in 1974, the South Dakota Act
has not been widely used.[149] Most major state actions have also involved Federal funding thus invok-

ing the requirements of NEPA rather than the state
law. Those actions that have required compliance
with SEPA have been relatively minor in scale and
the preparation of an "environmental assessment form"
type report (EAF) by the proposing agency has often
sufficed in satisfying the state requirements.

In fiscal 1975-76, for instance, there were only
a total of ten environmental documents (EAF's and
EIS's) prepared under SEPA:

TABLE 5.5	
TOTAL SEPA DOCUMENTS	
Fiscal 1975	
Transportation projects	1
Water management projects	5
Park improvement	1
Airport improvement	1
Physical modifications	1
Other	1
Total	10

The total caseload for fiscal 1976-77 was estimated
at even less than ten documents.

*Thus, the costs associated with SEPA have been
extremely small since 1974.* The Department of Environmental Protection, for example, has experienced
very little incremental workload as a result of the
law and review responsibilities under SEPA have generally been incorporated into the existing scheme of
operation. Likewise, the costs to project-originating agencies have been minimal when compared to the
expenses incurred in complying with NEPA. It is anticipated, however, that newly passed energy facility
siting regulations will result in increased EIS activity at the state level beginning July 1, 1979.[150]

TEXAS

In March 1972, the Governor's Interagency Council on Natural Resources and the Environment (ICNRE) adopted a "Policy for the Environment" for the State of Texas. This policy suggested that member agencies, as part of their planning process, should develop an environmental impact statement for the significant state-supported projects which do not include Federal funds. In 1975, the policy statement was supplemented by procedural suggestions and reissued as "The Environment; Policy--Guidelines and Procedures for Processing EIS's."

The statement of policy as well as the initial description of suggested agency responsibilities are virtual repetitions of the language of NEPA. The description of EIS content, however, takes on quite a different tone from the Federal mandate:

> The Council suggests that all member agencies shall:
> c. Include in project proposals significantly affecting the quality of the human environment a detailed statement by the responsible action agency on:
> (i) the expected <u>beneficial environmental impacts</u> of the proposed actions,
> (ii) any adverse, unavoidable environmental effects if the proposed actions are implemented,
> (iii) the alternative methods of accomplishing the proposed actions,
> (iv) the relationships between local, short-term uses of man's environment and the expected long-term effects on the environment,
> (v) any irreversible ecological changes or irretrievable commitments of resources which may be expected if the proposed actions are implemented,
> (vi) the approximate extent to which the proposed action will affect the supply and <u>conservation of energy</u> in the State as well as a statement of energy efficiency.

(Emphasis added)

Each participating state agency is asked to categorize projects or activities, in consultation with the Division of Planning Coordination, as follows:

Type I Projects - will always require EIS's
Type II Projects - may or may not require EIS's depending on the individual significance of the projects
Type III Projects - will not require EIS's

Type II projects are assessed to determine if they will have a significant effect through an initial evaluation process; data is first assembled for a "profile" of the economic, social, and environmental aspects of the planning area. The profile is then extended into the future to depict "without-project" conditions. Each alternative plan is analyzed in order to estimate the "with-project" effects.

If, after this analysis, it is decided that an EIS is required, the Office of the Governor, Division of Planning Coordination (DPC) coordinates the review of the EIS for the originating member agency; draft impact statements are distributed to the member agencies of ICNRE and those agencies having recognized interest, legal jurisdiction and/or technical competence in the applicable environmental matters. DPC then requests that these agencies review and comment on the draft EIS or take other necessary action to meet their responsibilities in the fields affected by the proposed actions (see Figure 5.3). After a thirty day review period, comments are returned to DPC where an attempt is made to resolve major issues presented in the review process. Problems which cannot be resolved among the agencies may be submitted to ICNRE for review.

It should be stressed that the policy, guidelines, and procedures adopted by ICNRE "neither requires nor mandates" its member agencies, but rather "suggests and solicits" the cooperation and coordination of its participants to appraise and improve the environmental effects of their activities and to develop new initiatives to abate environmental problems.

COST AND CASELOAD

With few exceptions, EIS's prepared under the State Policy also involve compliance with NEPA since most major projects involve Federal funds in some way. Thus, the incremental costs associated with

FIGURE 5.3
TEXAS EIS PROCESS

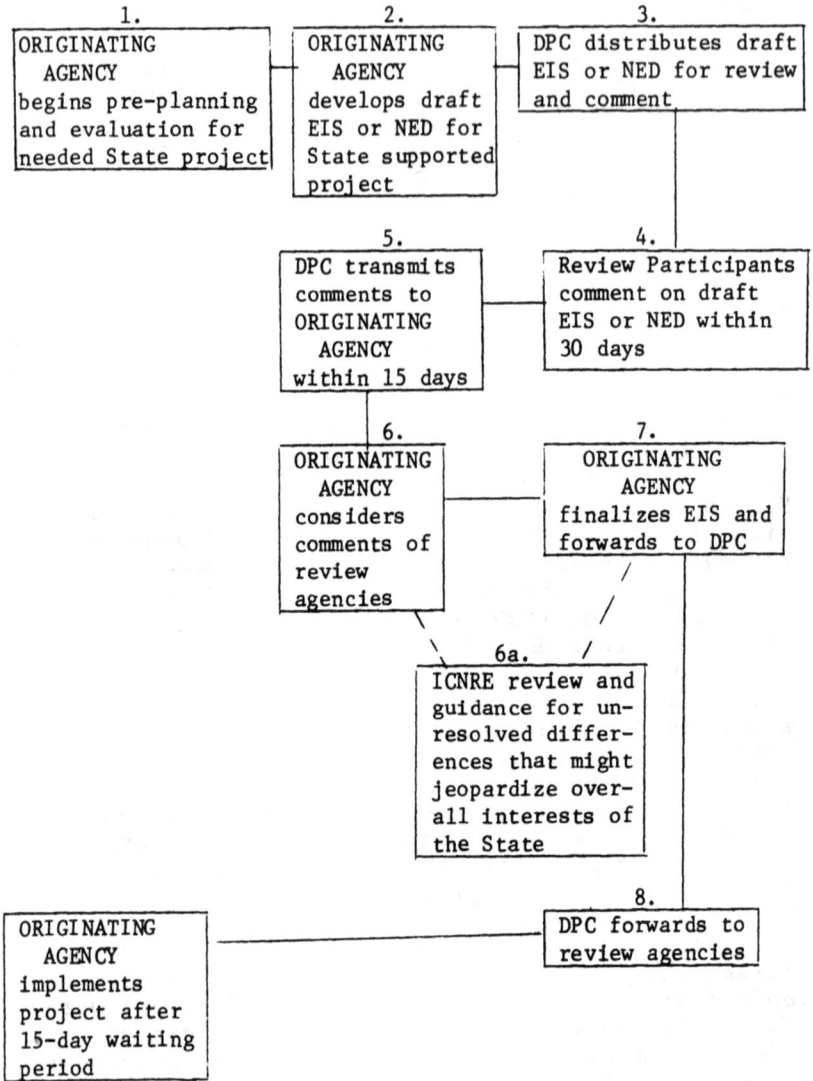

| 1. ORIGINATING AGENCY begins pre-planning and evaluation for needed State project | 2. ORIGINATING AGENCY develops draft EIS or NED for State supported project | 3. DPC distributes draft EIS or NED for review and comment |

| 5. DPC transmits comments to ORIGINATING AGENCY within 15 days | 4. Review Participants comment on draft EIS or NED within 30 days |

| 6. ORIGINATING AGENCY considers comments of review agencies | 7. ORIGINATING AGENCY finalizes EIS and forwards to DPC |

6a.
ICNRE review and guidance for un-resolved differ-ences that might jeopardize over-all interests of the State

ORIGINATING AGENCY implements project after 15-day waiting period

8. DPC forwards to review agencies

210

the state EIS requirement have been minimal.[151]

Since the state EIS policy is voluntary, few
project initiators have chosen to carry comprehensive environmental review past the stage of preliminary assessment and Negative Declaration. For
example, in 1975, there were forty Negative Declarations prepared as compared to only twenty-six preliminary assessments and EIS's. In 1977, these dropped to seven and four respectively. Indeed, the majority of environmental documents prepared are preliminary assessments by the Department of Highways
and Public Transportation which is already required
to submit assessments under NEPA.[152]

*Thus, the Texas EIS policy carries little weight
in the reality of project planning and review* and it
was expressed that the interaction of the various
state permit and license requirements supplies ample
consideration of environmental concerns.[153]

UTAH

A comprehensive environmental impact statement requirement was adopted by the State of Utah through Executive Order in August 1974. Although the "Executive Order on Environmental Quality" contained a statement of policy that was a virtual repetition of NEPA's Section 101, the definition of EIS applicability and content differed modestly from the Federal model:

> B. Pursuant to the guidelines which shall be adopted for the implementation of this Order, each agency of state government shall submit for every state action it proposes or is administratively responsible for which has the potential significantly to affect the environment, a statement on:
> (1) a description of the proposed action,
> (2) a description of the environment affected,
> (3) the probable impact of the proposed action on the environment,
> (4) mitigative measures included in the proposed action,
> (5) the relationship between local short-term uses of man's environment and the maintenance and enhancement of long-term productivity, and
> (6) any irreversible and irretrievable commitments of resources that would be involved in the proposed action should it be implemented. (Emphasis added)

An Environmental Coordinating Committee (ECC), composed of representatives of nineteen state agencies and divisions, was established to assist the existing "Economic and Physical Development Interdepartmental Coordination Group" in the review of EIS's and the promulgation of implementing guidelines.

"Guidelines for Implementation of the Executive Order on Environmental Quality" were promulgated by ECC soon after the Order was issued in 1974. Aside

from detailing the procedures for review set forth
in the Executive Order, the guidelines established
criteria for determining significant environmental
impact. This is important since both the power of
threshold determination and EIS adequacy are retained within the ECC:

> VIII. Criteria for Determining Significant En-
> vironmental Impact
> Significant environmental impact shall
be determined according to the relative magnitude of an effect and its probability of occur-
ring. Small effects which may by themselves be
insignificant, may have a cumulative effect
that is significant. Public reaction to a proposed action, irreversibility of that action,
duration of impact, and scope and stability of
affected ecosystems are some of the factors
that are relevant in determining significance.
As a general rule, if a proposed action may result in the following, it shall be considered
to have significant environmental impact:
> (1) Conflict with state or local master plans;
> (2) Relocation or disruption of population;
> (3) Encroachment on Municipal, State, or Na-
> tional Historic Sites, Parks, Monuments,
> or Recreational Sites;
> (4) Impact deemed to be significant by any
> public agency on big game winter range,
> wildlife management areas, game refuges,
> or habitat of rare or endangered species;
> (5) Impact deemed to be significant by any
> public agency on soils, bed rock strengths,
> and mineral resources;
> (6) Impact deemed to be significant by any
> public agency on air, noise, or water
> quality;
> (7) Impact deemed to be significant by any
> public agency on municipal water supplies
> and their watersheds;
> (8) Impact deemed to be significant by any
> public agency on a live stream bed, lake
> shore or aquatic environment;
> (9) Impact deemed to be significant to histor-
> ical or archaeological sites;
> (10) Impact deemed to be significant by any
> public agency on other environmental fea-
> tures deemed to be important by that agency.

The Office of the State Planning Coordinator was
designated as the agency responsible for the mechanics

of document dissemination and review. *However, the Executive Order set forth an innovative approach to review and enforcement:*

> If approval of a project is recommended by unanimous vote of the Environmental Coordinating Committee, it shall be considered approved without further action. If there is a divided vote on a project in the Environmental Coordinating Committee, the project shall be submitted to the Economic and Physical Development Inter-Departmental Coordination Group for consideration. If that group shall approve the project unanimously, it shall be considered approved for all purposes. If the project is disapproved by the Economic Physical Development Inter-Departmental Coordination Group, or is approved on a divided vote, the project shall be submitted to the Governor for final action. (Emphasis added)

Finally, all state agencies are required to respond to the initiating agency within thirty days of the initial request, and review comments are required from appropriate Federal and local agencies within forty-five days; if such comments are not forthcoming within these time limits, the process shall proceed without such approval. *Thus, the Utah system imposes strong review and enforcement powers but balances these with extremely distinct time limitations.*

COST AND CASELOAD

Although no attempts have been made to assess the costs of the state EIS program, it is possible to generate rough estimates. Since the bulk of technical assistance and review staff is provided by those agencies represented on the Environmental Coordinating Committee, an estimate of time expenditures by these agencies gives a good picture of EIS preparation, processing, and review costs. As many as thirty individuals are involved in the EIS process (both state and Federal) totaling between five and ten person-years per year.[154] Approximately 25 percent of this time is for secretarial staff at a salary range of $7,000-$10,000 per year; the remaining 75 percent of the time involves professional staff earning between $25,000 and $30,000 per year.[155] Thus, the cost range breaks down as follows:

Minimum EIS Cost per year (State & Federal)
```
5 person-years/yr.
   75% at $25,000/yr = $ 93,750
   25% at $ 7,000/yr = $  8,750
                       $102,500 per year
```

Maximum EIS Cost per year (State & Federal)
```
10 person-years/yr
   75% at $30,000/yr = $225,000
   25% at $10,000/yr = $ 25,000
                       $250,000 per year
```

In 1976, twenty-five environmental assessments were prepared pursuant to the State Executive Order. This represented a slight increase over the previous two years of implementation. *It is important to note, however, that no full environmental impact statements have yet been prepared under the Utah Order*--problems associated with projects (the bulk of which are proposed by state agencies rather than private parties) have thus far been resolved in the preliminary assessment stage.[156] Therefore, the costs associated exclusively with the state EIS requirement are probably slight.

VIRGINIA

Virginia's commitment to environmental protection and the wise use of natural resources was spelled out in Article XI of the revised State Constitution (1970). The Virginia Environmental Quality Act (VEQA), passed in 1972, incorporated this constitutional mandate into the Code of Virginia as a matter of policy. VEQA also created the Council on the Environment (COE) and assigned it broad responsibilities in coordination, management, and counsel--especially in the area of state permit procedures.[157] The Council is composed of ten unpaid members (primarily Chairpersons of State Boards) and an Administrator who also serves as Chairperson.

In 1973, a separate law was passed requiring the preparation of environmental impact statements on all statewide construction projects proposed by the executive branch of State government costing $100,000 or more. The law categorically excluded highway projects, however, since most, if not all, such projects are subject to the requirements of NEPA.[158] The Council on the Environment (COE) was assigned responsibility for coordination of the program. In response to this mandate (and existing responsibility under NEPA), an "EIS Coordinator" position was added to the Council staff in 1973. Although no enforcement powers were granted to the Council, the legislation made it clear where the ultimate authority resided in approving state projects:

> §10-17.110. Approval of Governor required for construction of facility. The State Comptroller shall not authorize payments of funds from the State treasury for a major State project unless such request is accompanied by the written approval of the Governor after his consideration of the comments of the Council on the environmental impact of such facility. Provided, however, this section shall not apply to funds appropriated by the General Assembly prior to June one, nineteen hundred seventy-three, or any reappropriation by the General Assembly of such funds.

The 1977 Session further expanded the scope of the EIS law to provide for a coordinated Council review of major projects proposed by the legislative and judicial as well as executive branches of State government:

§10-17.108. State agencies to submit environmental impact reports on major projects. All State agencies, boards, authorities and commissions or any branch of the State government shall prepare and submit a report to the Council on each major State project. Such reports shall include, but shall not be limited to, the following:
(1) The environmental impact of the major State project;
(2) Any adverse environmental effects which cannot be avoided if the major State project is undertaken;
(3) Measures proposed to minimize the impact of the major State project;
(4) Any alternatives to the proposed construction; and
(5) Any irreversible environmental changes which would be involved in the major State project. (Emphasis added)

The definition of what constituted a "major State project" was also expanded to include the *acquisition* of land and the *expansion* of existing facilities as well as the construction of new facilities:

(b) *"Major State project"* means the acquisition of land for any State facility construction, the construction of any facility or expansion of an existing facility which is hereafter undertaken by any State agency, board, commission, authority or any branch of the State government, including state-supported institutions of higher learning, which costs one hundred thousand dollars or more; provided, this term shall not apply to any highway or road construction or any part thereof. For the purposes of this chapter, authority shall not include any industrial development authority created pursuant to the provisions of chapter 33 of Title 15.1 of this Code or chapter 643, as amended, of the 1964 Acts of Assembly. Nor shall authority include any housing development or redevelopment authority established pursuant to

217

State law. For purposes of this chapter, branch of the State government shall not be construed to include any county, city or town of the Commonwealth.

Thus the criteria for threshold determination and categorical exemption are set forth in the legislation of Virginia's EIS requirement; most other states have looked to guidelines for clarification in these areas.

Guidelines and a Procedures Manual were initially published by the Council in 1973 and subsequently updated in 1976. The Guidelines established an abbreviated report, the "Preliminary Environmental Impact Statem" (PEIS), as a primary tool of impact assessment. Additional criteria for the contents of PEIS's were included in the Supplemental Guidelines provided by the State Water Control Board, Virginia Institute of Marine Science, State Air Pollution Control Board, and the Department of Conservation and Economic Development. The specific information requested in the supplemental guidelines is to be provided in addition to the data in the general guidelines whenever any project potentially affects the areas of concern of those agencies. If there is any doubt as to the potential impact of a project on any given state resource (e.g., water, air, marine resources, and forestry), the guidelines state that it is the responsibility of the agency proposing the project to contact the state agency responsible for its protection:

> It should be emphasized that the Council's intent is to avoid the necessity for a full EIS by providing for the internalization of the process so that all alternatives have been considered adequately and plans to minimize adverse environmental effects have been incorporated prior to the preparation of the PEIS. For this reason it is necessary that the sponsoring agency contact the review agencies prior to preparing the PEIS to assess the compatibility of the project with review agency goals, programs, regulations and standards and to determine what information they will require in order to evaluate the document.

The EIS process, as defined in the Procedures Manual, is depicted in Figure 5.4. COE receives and circulates copies of the documents to agencies on a preselected list with interest or special expertise.

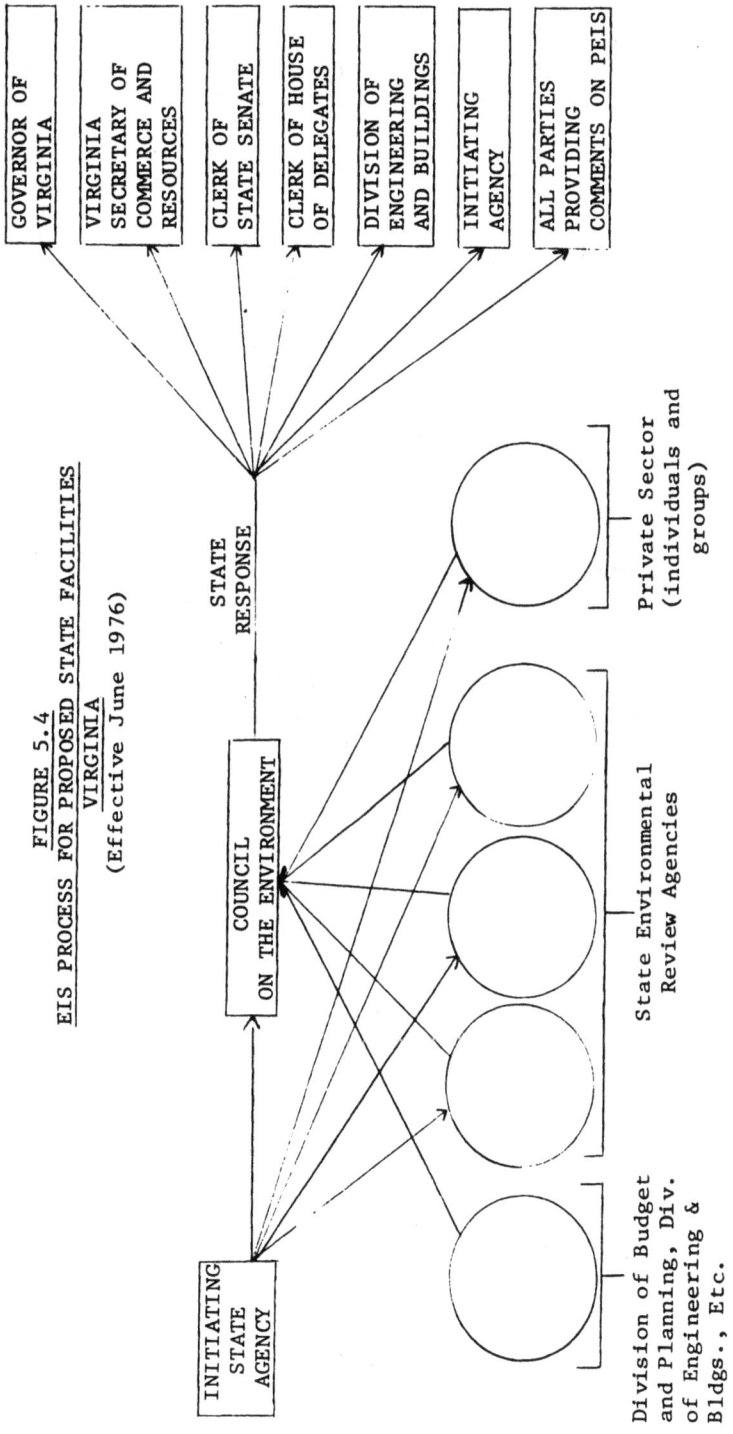

FIGURE 5.4
EIS PROCESS FOR PROPOSED STATE FACILITIES
VIRGINIA
(Effective June 1976)

GOVERNOR OF VIRGINIA

VIRGINIA SECRETARY OF COMMERCE AND RESOURCES

CLERK OF STATE SENATE

CLERK OF HOUSE OF DELEGATES

DIVISION OF ENGINEERING AND BUILDINGS

INITIATING AGENCY

ALL PARTIES PROVIDING COMMENTS ON PEIS

STATE RESPONSE

COUNCIL ON THE ENVIRONMENT

INITIATING STATE AGENCY

Private Sector (individuals and groups)

State Environmental Review Agencies

Division of Budget and Planning, Div. of Engineering & Bldgs., Etc.

Comments are returned to COE (generally within thirty days.) where a consolidated response is prepared and sent to a variety of interested parties in addition to the initiating agency. Public participation is encouraged through open Council meetings, communications with conservation groups, and the monthly publication of currently available EIS's.[159] Procedures for the implementation of the EIS law are currently in the process of revision and should be in place by late 1978.[160]

Since capital outlay requests by many state agencies are considerable, the Council requires a "Form of Intent" be submitted in January of every odd-numbered year (to coincide with Virginia's budget preparation schedule) by each state agency proposing projects for which funds will be requested for the subsequent biennium. *Through this device, the Council can anticipate the number of PEIS's to be submitted, for what projects, and when to expect them.* This information then becomes available to the review agencies as well as to all other interested persons. The primary benefit is that meaningful interactions can begin between the agency sponsoring the projects and all interested parties early in the planning process and, therefore, modifications can be made to the proposed facilities, if necessary, prior to the preparation of the PEIS.

COST AND CASELOAD

Appropriations for the Council on the Environment were $83,895 in fiscal 1974-75 and $85,220 in fiscal 1975-76. These figures were reflective of the following level of staffing:

Administrator
Environmental Impact Statement Coordinator (1)
Confidential Secretary (to Administrator)
Clerk/Stenographer (1)
TOTAL (4)

In 1976, the Council received the added responsibility of producing a joint environmental agencies budget in response to the State's conversion to *program budgeting.* Requested appropriations jumped to $176,740 in 1976-77 and $181,770 in 1977-78 and the requested level of staffing included the following:

Administrator
Assistant Administrator
Environmental Impact Statement Coordinators (2)

Budget and Management Analyst
Confidential Secretary (to Administrator)
Clerk-Stenographers (2)
TOTAL (8)

Actual appropriations, however, included only
two new positions (for a total of six) and a total
of approximately $117,000 for each of the two years.
Requests for the 1978-80 biennium total $269,335 or
about $135,000 per year.[161] *This constitutes only
.26 percent of the overall environmental budget*
(see Figure 5.5).
It must be stressed again, however, that the
coordination of the state EIS law represents only a
fraction of the overall responsibilities of the Council as established under VEQA. In 1975, for example,
COE processed and reviewed eighty-nine PEIS's under
the State law as well as sixty-five EIS's under NEPA.
In fiscal 1976, caseload under the state EIS requirement jumped to 120 PEIS's, the bulk being for prisons,
parks, and college facilities.[162] Thus, the cost directly ·attributable to the state EIS law is only a
part of the larger EIS program and much ·less than
the overall Council program budget. It is estimated
that an average of twelve person-hours is spent in
the review and processing of each EIS within COE.[163]
Assuming an average annual salary of $19,000[164] and
120 PEIS's per year, this amounts to about one person-year of $19,000 actually being spent on state
EIS program coordination.
The costs of document preparation under the
state program are much more difficult to assess
since they do not show up on any program or budget
line item. However, since the primary objective of
the program is to establish project compatibility
with review agency goals early on, the cost of actual
documentation has been minimal; on the average, PEIS's
are five to six pages long and take only about eight
hours to prepare.[165] A virtual lack of litigation
associated with the law has also contributed to the
generally informal implementation style. *Thus,
Virginia's EIS law has differed from NEPA to the extent that controversy has not yet forced agencies to
provide exhaustive formal documentation of environmental impacts.*

FIGURE 5.5

VIRGINIA

AGENCY TOTAL (GENERAL + SPECIAL) FUND REQUESTS

1978-80 BIENNIUM

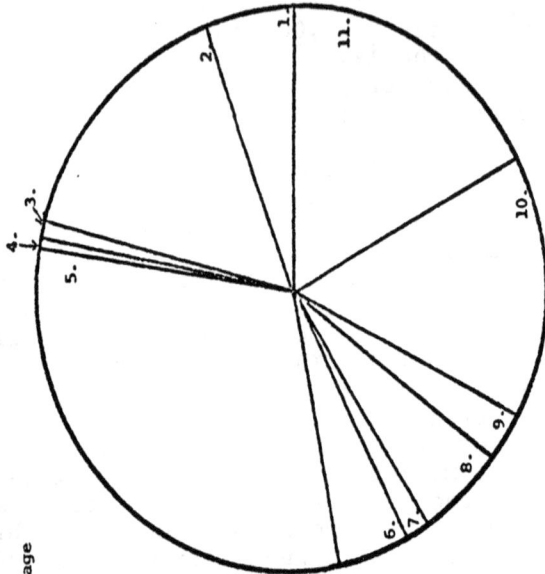

Agency	Dollar Amount	Percentage
1.Air Pollution Control Board	5,748,970	5.61%
2.Commission of Game and Inland Fisheries	15,683,870	15.31%
3.Commission of Outdoor Recreation	919,695	.90%
4.Council on the Environment	269,335	.26%
5.Dept. of Conservation & Economic Development	31,297,395	30.56%
6.Dept. of Health	4,582,690	4.47%
7.Historic Landmarks Commission	1,492,345	1.46%
8.Marine Resources Commission	5,605,610	5.47%
9.Soil and Water Conservation Commission	2,821,670	2.76%
10.State Water Control Board	15,266,468	14.90%
11.Virginia Institute of Marine Science	18,751,480	18.30%
TOTAL	$102,439,528	100%

WISCONSIN

The Wisconsin State Legislature passed two significant environmental impact laws which became effective in April 1972. The first, Chapter 274, Laws
of 1971, is known as the Wisconsin Environmental
Policy Act (WEPA). This law requires all "agencies
of the State" to prepare environmental impact statements on proposals for legislation and for other major actions (including management and regulatory
functions of agencies) significantly affecting the
quality of the human environment. While the language
of WEPA is very similar to that of NEPA in both policy and action-forcing procedure, WEPA contains two
significant additional requirements:

1. Each impact statement shall also contain
   details of the beneficial aspects of the
   proposed project, both short term and long
   term, and an analysis of the economic advantages of the proposal. The 1973 Legislature amended the Act to require agencies
   to assess the economic disadvantages of
   each proposal as well; and
2. A public hearing shall be held on each
   proposal (other than legislation) for which
   an EIS is prepared.

The second Act, Chapter 273, Laws of 1971, created Section 23.11(5), Wisconsin statutes. This law
provides the mechanism for the Department of Natural
Resources to request an *impact report* from applicants
seeking Departmental permits or statutory approval
for various projects (both public and private), provided the area affected exceeds forty acres or if
the estimated cost of the project exceeds $25,000.
If, after reviewing this report, DNR makes a determination that its regulatory action will be a "major action significantly affecting the quality of
the human environment," then the Department prepares
an EIS pursuant to WEPA. If the ultimate regulatory
action is not considered to be significant, the impact report is kept in the project file for reference.[166]

Thus, an applicant's impact report provides essential background information about a proposed project and forms the basis of the impact statement prepared by the department.

Shortly after WEPA became effective, the Governor issued a set of guidelines to implement the law. It soon became apparent, however, that the language of the law was broad and subject to a variety of interpretations by state agencies and the public.[167] Thus, in November 1972, the Governor established an Inter-Agency Environmental Impact Statement Coordinating Committee to refine the guidelines and to establish a more uniform approach for implementing the Act. A second set of guidelines was issued through Executive Order in December 1973.

Governor's Executive Order #69 established guidelines to be used by state agencies for implementing WEPA (see Figure 5.6). The guidelines require that the agency prepare a preliminary environmental report, followed by an environmental impact statement and a public hearing. Under these guidelines, the preliminary environmental report (PER) is circulated for a forty-five day review to state, Federal, and local agencies with expertise or concerns related to the project, and it is also made available to the public. Comments and questions submitted to the Department on the PER are used to develop an EIS. The EIS is circulated to the commenting agencies and the public for a thirty day review. A hearing is then held to receive the views of the public on the Department's environmental impact statement. Following the public hearing, the Department formulates a conclusion on its decision on the adequacy of the EIS. This decision is circulated to commenting agencies and the public. *Only upon completion of the EIS process are any required regulatory hearings scheduled.*

The revised guidelines also set up several specific requirements for state agencies. Each agency must, for example, assign its WEPA responsibilities to one individual known as a "WEPA Coordinator." (All Coordinators serve on the Interagency WEPA Coordinating Committee and may be contacted by other agencies or the public regarding environmental review matters of their agency). Each agency must also:

1. develop "screening worksheets" (assessment formats for evaluating the environmental significance of proposed projects);
2. develop specific environmental impact

FIGURE 5.6
ENVIRONMENTAL REVIEW PROCESS
Under The Wisconsin Environmental Policy Act

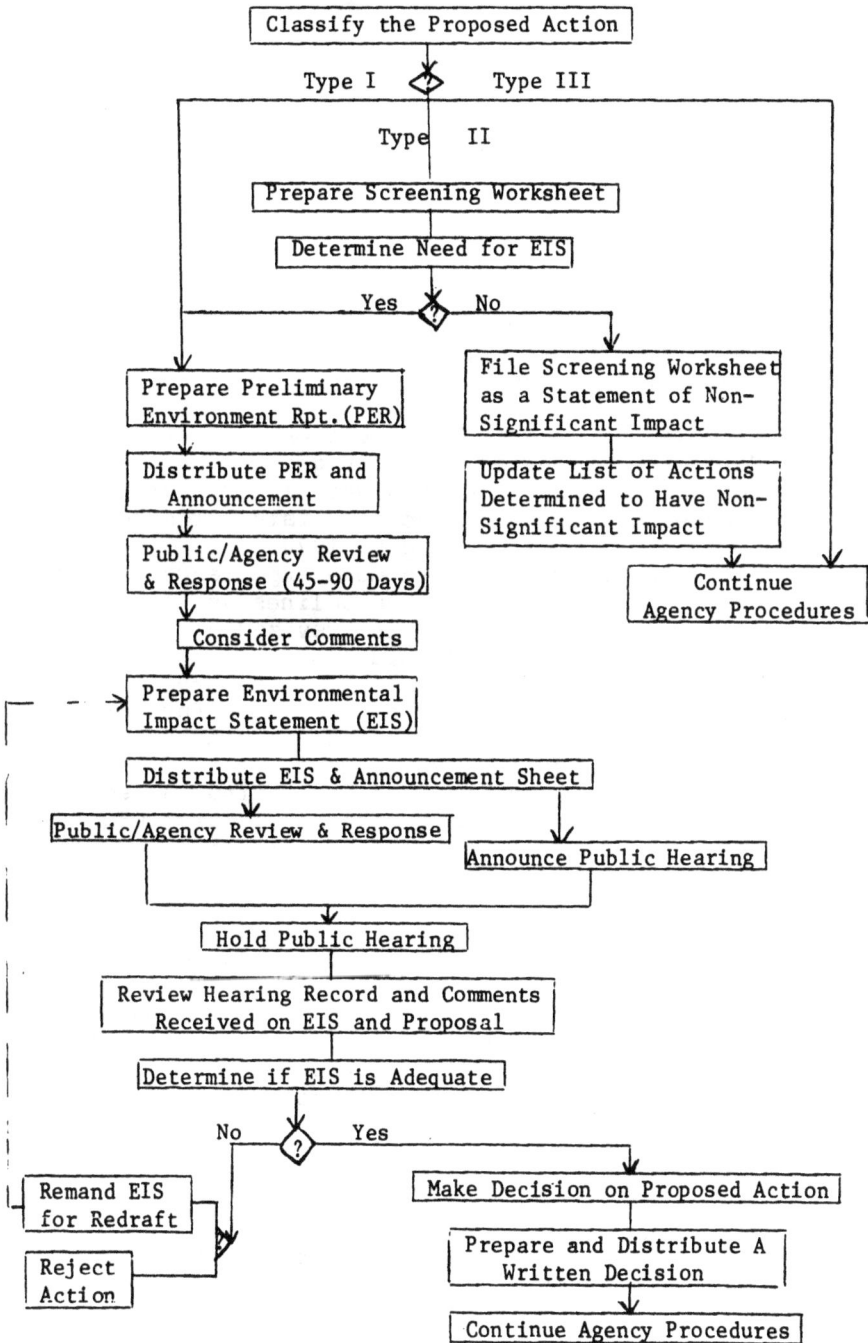

```
                 ┌─────────────────────────────────┐
                 │   Classify the Proposed Action   │
                 └─────────────────────────────────┘

           Type I      ◇        Type III

                   Type    II

            ┌───────────────────────────────┐
            │  Prepare Screening Worksheet   │
            └───────────────────────────────┘

              ┌───────────────────────────┐
              │  Determine Need for EIS    │
              └───────────────────────────┘

                   Yes   ◇?   No

┌──────────────────────────┐    ┌────────────────────────────┐
│  Prepare Preliminary      │    │ File Screening Worksheet    │
│  Environment Rpt.(PER)    │    │ as a Statement of Non-      │
└──────────────────────────┘    │ Significant Impact          │
                                 └────────────────────────────┘
┌──────────────────────────┐    ┌────────────────────────────┐
│  Distribute PER and       │    │ Update List of Actions      │
│  Announcement             │    │ Determined to Have Non-     │
└──────────────────────────┘    │ Significant Impact          │
                                 └────────────────────────────┘
┌──────────────────────────┐
│  Public/Agency Review     │              ┌──────────────────┐
│  & Response (45-90 Days)  │              │    Continue       │
└──────────────────────────┘              │ Agency Procedures │
                                           └──────────────────┘
       ┌──────────────────────┐
       │  Consider Comments    │
       └──────────────────────┘

   ┌──────────────────────────┐
─→ │  Prepare Environmental    │
   │  Impact Statement (EIS)   │
   └──────────────────────────┘

   ┌────────────────────────────────────────┐
   │  Distribute EIS & Announcement Sheet     │
   └────────────────────────────────────────┘

┌──────────────────────────────────┐
│ Public/Agency Review & Response   │   ┌──────────────────────────┐
└──────────────────────────────────┘   │ Announce Public Hearing   │
                                        └──────────────────────────┘

         ┌──────────────────────────┐
         │   Hold Public Hearing     │
         └──────────────────────────┘

   ┌────────────────────────────────────────┐
   │ Review Hearing Record and Comments       │
   │ Received on EIS and Proposal             │
   └────────────────────────────────────────┘

   ┌────────────────────────────────────────┐
   │ Determine if EIS is Adequate             │
   └────────────────────────────────────────┘

          No    ◇?    Yes

┌──────────────────┐      ┌────────────────────────────────────┐
│ Remand EIS        │      │ Make Decision on Proposed Action    │
│ for Redraft       │      └────────────────────────────────────┘
└──────────────────┘      ┌────────────────────────────────────┐
┌──────────────────┐      │ Prepare and Distribute A            │
│ Reject            │      │ Written Decision                    │
│ Action            │      └────────────────────────────────────┘
└──────────────────┘      ┌────────────────────────────────────┐
                          │ Continue Agency Procedures          │
                          └────────────────────────────────────┘
```

225

assessment procedures; and
3.   develop a list of available expertise
      within the agency indicating specialists
      for impact statement preparation and re-
      view.

Perhaps the most important aspect of these requirements, however, is the *categorization of agency
actions* to identify those "major actions" with potentially "significant effects." By April 1974, all
state agencies had categorized their actions according
to the following three types:

1.   Type I actions which always require an EIS;
2.   Type II actions which may or may not require
      an EIS (depending on individual project
      significance); and
3.   Type III actions which never require an EIS.

Upon completion, these agency action lists were circulated to other agencies, citizen groups and the public.
Many of the public's comments were incorporated into
a redrafted set of agency action lists. The revised
version was published in August 1974.
      As the agencies gained more experience with WEPA,
it became apparent that the guidelines and agency action lists needed to be further revised and updated.
To this end, both revised guidelines and the agency
action lists were reissued by Executive Order in February 1976. The new guidelines provided further interpretation of the Act including definitions of "major" and "significant." Refinements in the impact
statement process are expected to continue as societal
priorities change. In fact, the new Federal CEQ Regulations for NEPA will be used in guideline revision
sometime in late 1978.[168]
      *The Wisconsin Environmental Policy Act did not
establish enforcement powers in any single agency.*
However, all state agencies are required to consult
with other agencies during the development of an EIS.
State agencies are also required to submit copies of
their statements to the Governor, the the Department
of Natural Resources, and to the public for review
and comment; the Inter-Agency Coordinating Committee
serves only as a sounding board for questions relating to procedural requirements of the Act.

COST AND CASELOAD

In 1975, the WEPA Coordinating Committee sur-
veyed its members to estimate the direct cost of the

state EIS law.[169] The figures derived from that
survey are still considered good estimates of WEPA-
related costs and caseload.[170]

A "WEPA Manpower and Funding Questionnaire" was
completed by the thirty state agencies affected by
WEPA. In computing their manpower and funding requirements, the agencies made every effort to limit
manpower and cost estimates to those attributable
directly to WEPA requirements. WEPA-related activities can be broken down into three categories:

1. Type II screenings;
2. Preliminary Environmental Reports (PER's)
   and EIS's initiated; and
3. PER's and EIS's reviewed.

The estimates do not include activities such as data
collection, engineering studies, and hearings which
are required by other laws, procedures, or regulations.

From July 1, 1974 to June 30, 1975, fourteen
of the thirty agencies affected by WEPA had experienced some WEPA-related activities. The estimated
total cost of this WEPA-related activity was $605,584
for the twelve month period. Of the fourteen state
agencies that experienced WEPA-related activities,
four agencies--Department of Natural Resources, Department of Transportation, Public Service Commission,
and University of Wisconsin--were responsible for
95.2 percent of all Type II screenings carried out,
89.3 percent of all PER's and EIS's initiated and
36.9 percent of all PER's and EIS's reviewed. The
total cost of WEPA-related activity for these four
agencies was $579,310, or 95.2 percent of the total
cost of all WEPA-related activities for all state
agencies (see Table 5.6).

Worksheets for the Type II screenings carried
out by these four major agencies were prepared primarily by in-house staff with assistance from other
agencies. The PER and EIS activities were handled
by a combination of in-house staff, other agencies'
staffs, university researchers, private consultants
and other governing units. In response to a question concerning the agency's preference for meeting
its WEPA staffing needs, the four major agencies preferred to have their own individual environmental
review staffs.

The remaining ten agencies[171] that experienced
WEPA-related activities from July 1, 1974 to June
31, 1975 were responsible for 4.8 percent of all
Type II screening carried, 10.7 percent of all PER's

TABLE 5.6
WEPA-RELATED ACTIVITIES AND COSTS FOR THE PERIOD
July 1, 1974–June 30, 1975[1]

Category	Agency	$, % of Total $	# of Type II Screenings, % of all Type II Screenings	# of PERs & EISs initiated, % of all PERs & EISs initiated	# PERs & EISs reviewed; % of all PERs & EISs reviewed
Major Agencies	Dept. of Natural Resources*	$301,390, 49.8%	1407, 89.4%	21, 32.3%	53, 21%
	Univ. of Wisconsin	$165,000, 27.2%	71, 4.5%	30, 46.2%	30, 11.9%
	Public Service Comm.*	$94,920, 15.7%	16, 1%	5, 7.7%	0, 0%
	Dept. of Transportation	$18,000, 3%	4, .3%	2, 3.1%	10, 4%
SUBTOTAL		$579,310, 95.7%	1498, 95.2%	58, 89.3%	93, 36.9%
TOTAL		$605,584, 100%	1573, 100%	65, 100%	252, 100%

*These agencies also have major fiscal commitments under NEPA and other environmental laws which are not reflected here.

[1]The hidden costs of WEPA include the personnel and administrative costs related to the position of "WEPA Coordinator" in each of the state agencies affected by WEPA and to the meetings and activities of the Interagency WEPA Coordinating Committee. An attempt was made to estimate this cost only in the case of the Department of Administration which has provided services to the Executive Office and staff support to the Interagency Committee.

and EIS's initiated, 63.1 percent of all PER's and EIS's reviewed. The estimated total cost of WEPA-related activities for these ten agencies was $26,274, or 4.3 percent of the total cost of all WEPA-related activities for all state agencies. Type II screenings and PER/EIS activities were handled primarily by in-house staff. The ten agencies expressed concern about fully complying with WEPA, however, because they did not have technical in-house staff capabilities; many requested access to a central consulting staff that could be loaned to these agencies as needed.

Another sixteen state agencies[172] participated in the deliberations of the Interagency WEPA Coordinating Committee but were not involved in the WEPA review activities and had no WEPA-related costs.

No budget allocations were provided to implement WEPA. *Agencies were in essence required to initiate environmental review with existing resources by shifting operating budgets and personnel from other programs.* As a result, many agencies have suffered from inadequate staff resources to fully comply with WEPA.[173] The agencies with the most limited staff capabilities or the most infrequent EIS preparation workload indicated a desire for the establishment of a centrally located staff of "environmental analysts" who could be loaned to assist agencies in fulfilling their WEPA requirements.

Some agencies, however, have developed special capabilities to meet the spirit and intent of WEPA. The Department of Natural Resources, for example, established a "Bureau of Environmental Impact" in 1971 to coordinate activities related to the Federal and State EIS laws. Although the BEI consisted initially of only two staff persons, it has since expanded into a sizable collection of interdisciplinary professionals. In fiscal 1976 this included:

- Director
- Assistant to the Director
- Chief - Biological Investigations
- Chief - Watershed Projects
- Chief - EIS Preparation
     with 6 Environmental Specialists
- Chief - EIS Review and Analysis
     with 2 Environmental Specialists

As of the beginning of fiscal 1978, the Department is authorized to fill the following additional positions:

- 2 Environmental Specialists for EIS
  Preparation Section
- 1 Clerical person for BEI
- 1 Administrative Assistant
- 1 Engineering Technician
- 3-person Statewide Environmental Impact
  Field Team to assist the Department's
  district staff in data collection.[174]

The Bureau staff is frequently called upon to assist other state agencies and the Bureau Director was appointed by the Governor in 1972 to co-chair the Inter-Agency Environmental Impact Coordinating Committee which provides guidance to all other agencies in administering WEPA.

Each month, approximately 125 environmental impact assessments are prepared on the Department's management and regulatory activities to determine the significance of their impact. The Department annually completes eighteen to twenty environmental impact statements and has an additional fifteen statements in various stages of completion. Each year the staff reviews about 150 environmental impact statements prepared by other Federal and state agencies.[175]

The review and preparation of environmental impact statements by Department staff is a team effort. Some of the more complex projects have required input from fifteen to twenty staff members. Cost of data collection, analysis of impacts and the preparation of an environmental impact statement varies with the project and has ranged from $6,000 to more than $150,000.[176] A law passed in 1975 (Section 23.40) required the Department of Natural Resources to charge a fee of .05% (factor .0005) of the estimated cost of the project for which the applicant seeks a permit, license, or approval when DNR determines it must prepare an EIS. However, revenues derived from this fee would have gone to the general fund with no direct relief for WEPA activities. As a result, the law was repealed the following year despite anticipated revenues of $1.5 million or more annually.[177] The law was revised in 1977 to allow the Department to charge only administrative costs of EIS preparation for private projects. In May 1978, the Governor signed a bill changing Section 23.40 for a second time. Under this revision of the EIS fee law, *the Department must now charge the full cost for preparation of EIS's on private projects*. Rules have been drafted to provide the mechanics for determining and collecting the EIS fee and the Department expects

the cost of the new positions to be offset by EIS
fees collected on major projects.[178]

Notes

1. In fact, the proposed Federal CEQ Regulations for NEPA implementation (43 Federal Register, 25230, issued June 9, 1978) contain several provisions which are taken directly from various State Acts. For more detail, see Enk, Gordon and Stuart Hart, Beyond NEPA Revisited: Directions in Environmental Impact Review, The Institute on Man and Science, Rensselaerville,NY, March 1978.

2. We are indebted to Davis Cherington, Research Associate at The Institute, for his work in developing the "EIS Cost Accounting System."

3. The California Environmental Quality Act; An Evaluation Emphasizing its Impact upon California Cities and Counties with Recommendations for Improving its Effectiveness, Volume III-Discussion of Costs and Delays Related to Environmental Review, prepared by Environmental Analysis Sytems, Inc., San Diego, CA, November 1975, p. 4.

4. Thaddeus Tryzna, Environmental Impact Requirements in the States: NEPA's Offspring, prepared for the Office of Research and Development, U.S. EPA, April 1974, p. 19.

5. Kenneth Pearlman, "Little NEPA's: The Impact Statement at the State Level," Department of City and Regional Planning, The Ohio State University, prepared for presentation to the American Institute of Planners, 1975, p. 4.

6. Ibid., p. 5.

7. Tryzna (1974), op. cit., p. 21.

8. The problems inherent in cost-benefit analysis are legend. For further discussion, see: Ezra Misham, Cost-Benefit Analysis, Praeger, NY, 1971; Charles Howe, Benefit-Cost Analysis for Water System Planning, AGU, Washington, 1971; Jerome Rothenberg, "Cost-Benefit Analysis: A Methodological Exposition," in Handbook of Evaluative Research, Elmer Strueming

and Marcia Guttentag, eds., Sage Publications, Beverly Hills, CA, 1975.

9.   Tryzna (1974), op. cit., p. 22.

10.   Thaddeus Tryzna and Arthur Jokela, California Environmental Quality Act; Innovation in State and Local Decisionmaking, prepared for the Office of Research and Development, U.S. EPA, October, 1974, p. 8.

11.   For a more in-depth discussion of NEPA's evolution, see Richard Liroff, A National Policy for the Environment: NEPA and its Aftermath, Indiana University Press, Bloomington, 1976, and Stuart Hart and Gordon Enk, Assessing the Benefits Derived from the Environmental Impact Statement Process, The Institute on Man and Science, Rensselaerville, NY (forthcoming).

12.   52 U.S.C. Section 1857 et seq.

13.   For a more complete discussion of NEPA's evolution and the concepts of "procedural" and "substantive," see Stuart Hart and Gordon Enk, Assessing the Benefits Derived from the Environmental Impact Statement Process, The Institute on Man and Science, Rensselaerville, NY (forthcoming).

14.   P. Heffernan and R. Corwin (eds.), Environmental Impact Assessment, Freeman, Cooper & Co., San Francisco, CA, 1975, p. 191.

15.   Environmental Impact Statements, A Report of the Commission on Federal Paperwork, Washington, D.C., February 25, 1977, p. 3.

16.   California Environmental Quality Act Litigation Study, prepared by the State Office of Planning and Research, April 1976.

17.   Extracted from questionnaire data.

18.   Hawaii, Indiana, Maryland, Massachusetts, Minnesota, Montana, North Carolina, and Wisconsin.

19.   Connecticut, Michigan, New Jersey, New York, South Dakota, Texas, Utah, and Virginia.

20.   Connecticut, California, Hawaii, Maryland, Massachusetts, Michigan, Minnesota, New Jersey, New York, North Carolina, Virginia, and Washington.

21.   Extracted from questionnaire data.

22.   See discussion of New York's EIS program in Chapter 4 of this report.

23.   While lack of appropriations and resultant staff "borrowing" are genuinely problems in the implementation of EIS requirements, it must also be recognized that it is rare that legislatures ever appropriate sufficient funds for any new program. Thus, the problem is much larger than environmental review and has more to do with the administration of state government in general.

234

24. Again, the reader is referred to the New York section in Chapter 4 for the details of these calculations.

25. We are indebted to Research Consultant Kate Troll for the synthesis of this information dealing with "units of impact."

26. This is especially true in the "pioneer" states of California and Washington. For a discussion of this aspect, see "The California Environmental Quality Act: The Integration of CEQA and Local Planning Practices," prepared by the Public Policy Research Organization, University of California at Irvine, January 1976.

27. Personal communication, William Blackmer, Assistant Chief, Office of Environmental Planning, CalTrans, State of California, 11/17/77.

28. Letter from Wayne Wetzel, Environmental Coordinator, Montana Department of Natural Resources and Conservation, May 30, 1978.

29. California Environmental Quality Act Litigation Study (1976), op. cit.

30. This estimation was substantiated by formal written material from Minnesota and a letter from the respondent in Montana's Department of Natural Resources and Conservation.

31. Indeed, this is the least optimistic estimate possible since it is based on figures from the Federal and California experience where the most litigation and delay has taken place.

32. Other factors such as the timing of EIS preparation, special site considerations, or litigation may drive certain individual project EIS costs much higher than the .5% average figure.

33. The Environmental Impact Statement--It Seldom Causes Long Project Delays but Could be More Useful if Prepared Earlier, a Report to the Committee on Environment and Public Works by the U.S. General Accounting Office, August 9, 1977, p. iii.

34. Ibid.

35. Ibid., p. 5.

36. Ibid., p. 6.

37. Personal communication, Robert Remen, Office of Planning and Research, California, 11/17/77.

38. Tryzna (1974), op. cit., p. 24.

39. Ibid., pp. 25-26.

40. A recent study by the Syracuse-Onondaga County Planning Agency (SOCPA) and funded jointly by the New York State Department of Environmental Conservation and SOCPA has shown this to be the case.

41. Supra., note 15.

42. Ibid., p. 51.

43. cf., "Implementation of the Minnesota Environmental Policy Act," prepared for the Minnesota Environmental Quality Council by Hayworth and Anderson, Inc., 1974.

44. <u>Loveless vs. Yantis</u>, 82 Wash. 2d 754, 513 P. zd. 1023, 1029 (Wash., 1973).

45. "Environmental Impact Statement Preparation: Is a Central Office the Answer?" A report to the Legislature by the Minnesota State Planning Agency/ Environmental Quality Board, February 15, 1978, p.13.

46. Personal communication, William Hicks, Chief Planner, Massachusetts Office of Environmental Affairs, 12/19/77.

47. cf. Steven Perlmutter, <u>The Montana Environmental Policy Act: The First Five Years</u>, Montana State Legislature/Environmental Quality Council, Helena, Montana, November 29, 1976.

48. Tryzna and Jokela (1974), <u>op. cit.</u>, p. 12.

49. Ibid, p. 1.

50. 8 Cal. 3d 247, 502 P. 2d. 1049, 104 Cal. Rptr. 761 (1972).

51. Tryzna and Jokela (1974), <u>op. cit.</u>, pp. 9-10.

52. Ibid., p. 10.

53. Ibid., p. 10.

54. Personal communication, Norman Hill, Assistant to the Secretary, The Resources Agency of California, 11/18/77.

55. Personal communication, Alcides Freitas, Environmental Coordinator, Environmental Impact Section, Sacramento County, CA, 11/18/77.

56. <u>The California Environmental Quality Act;</u> An Evaluation Emphasizing its Impact Upon California Cities and Counties with Recommendations for Improving its Effectiveness, prepared by Environmental Analysis Systems, Inc., San Diego, CA. November 1975, three Volumes (hereafter cited as "CEQA; An Evaluation" followed by the Volume number).

57. CEQA, An Evaluation, Volume I, pp. 44-45.

58. CEQA, An Evaluation, Volume III, p. 17.

59. CEQA, An Evaluation, Volume I, p. 40.

60. Personal communication, Don Meixner, California Department of Water Resources, 11/17/77.

61. Personal communication, William Blackmer, Chief, Office of Environmental Planning, CalTrans, State of California, 11/17/77.

62. Figures provided by William Blackmer (see note 61).

63. Personal communication, Robert Remen, Office of Planning and Research, California, 11/17/77.

64. Personal communication, Tom Elwell, Environmental Review Section, Washington Department of Ecology, 11/21/77.

65. Pearlman, Kenneth (1975), op. cit., p. 2.
66. Personal communication, Tom Elwell (see note 64).
67. "Implementation of the State Environmental Policy Act: A Report to the Legislature," prepared by the Washington (State) Office of Program Planning and Fiscal Management, November 24, 1976.
68. James Williams of the Washington State Association of Counties provided the preliminary figures found in the report.
69. Personal communication, Tom Elwell (see note 64).
70. Personal communication, Bernie Chaplin, Project Planning Supervisor, Washington Department of Transportation, 11/22/77.
71. Tryzna (1974), op. cit., p. 14.
72. Personal communication, Thayer Broili, North Carolina Department of Natural and Economic Resources, 4/25/78.
73. "Guidelines for the Implementation of the Environmental Policy Act of 1971," issued February 1972; revised February 1976 by the Department of Administration.
74. Roe, Charles. "The North Carolina Environmental Policy Act: Neglected Planning Tool," paper submitted to the Institute of Government, Chapel Hill, North Carolina, April 1975, p. 9.
75. Ibid., p. 6.
76. Chapter 25, N.C. Administrative Code.
77. The cost information is based on extensive personal communications with Keith Whitenight and Thayer Broili, both of the Department of Natural and Economic Resources. The bulk of the research was conducted by Research Consultants Katherine Troll, Joanne Polayes, and Mary Jo Waits during the summer of 1976.
78. Personal communication, Byron O'Quinn, Environmental Planning Engineer, N.C. Department of Transportation, 8/76.
79. Personal communication, William Hicks, Chief Planner, Massachusetts Office of Environmental Affairs, 12/19/77.
80. New Regulations issued in May 1978 effectively eliminated this two-system process.
81. Personal communication, William Hicks (see note 79).
82. Personal communication, Richard Bates, Massachusetts Department of Environmental Quality Engineering, 1/26/78.
83. Ibid.
84. Ibid.

85. Personal communication, John Hurley, Environmental Engineer, Massachusetts Department of Public Works, 12/19/77.

86. Letter from John Hurley, 6/28/78 (see note 85).

87. The information is again based on extensive personal interviews conducted by Research Consultants Katherine Troll, Joanne Polayes, and Mary Jo Waits.

88. Personal communication, William Hicks (see note 81).

89. Ibid.

90. Ibid.

91. Letter from Raymond Ghelardi, Massachusetts Office of Environmental Affairs 5/23/78.

92. Personal communication, William Hicks (see note 88).

93. Polayes, Joanne. "Environmental Policy Acts and the Public," a working paper of The Institute on Man and Science, 1976, p. 16.

94. "Implementation of the Minnesota Environmental Policy Act," prepared for the Minnesota Environmental Quality Council by Hayworth and Anderson, Inc., 1974.

95. Ibid., p. 3.

96. "Environmental Impact Statement Preparation: Is a Central Office the Answer?" A report to the Legislature by the Minnesota State Planning Agency/Environmental Quality Board, February 15, 1978.

97. Ibid., p. A-2.

98. Ibid., p. G-1.

99. The bulk of the research on the individual agency costs was again conducted by Research Consultants Katherine Troll, Joanne Polayes, and Mary Jo Waits during the summer of 1976.

100. Personal communication, Robert Herbst, Minnesota Department of Natural Resources, 8/76.

101. "Rules and Regulations for Environmental Impact Statements." (6 MCAR Sect. 3.021).

102. Extracted from questionnaire information.

103. Personal communication, Terence Curran, Director, Office of Environmental Analysis, New York Department of Environmental Conservation, 12/8/77.

104. State Environmental Quality Review Law; Local Assistance Program, p. 1.

105. Ibid., p. 2.

106. Ibid., p. 2.

107. Staff time estimates and all salary levels provided by Pat Grady of the Office of Environmental Analysis.

108. Personal communication, Keith Smith, En-

vironmental Impact Section, N.Y.S. Department of
Transportation, 12/8/77.
109. Ibid.
110. Personal communication, Bernie Cobb, Division of the Budget, N.Y.S. Department of Transportation, 12/8/77.
111. Ibid.
112. Personal communication, Peter Buttner, Director of Environmental Management, N.Y.S. Parks
and Recreation, 12/9/77.
113. Ibid.
114. Ibid.
115. All figures based on estimates by Peter
Buttner (see note 112).
116. Ibid.
117. Personal communication, Michael Morandi,
Office of Environmental Analysis, 8/31/78.
118. Latham, Janis. "Connecticut Environmental
Policy Act (CEPA): Problems with Implementation;
Proposals for Revision," Office of Legislative Research, October, 1975, pp. 4-5.
119. Letter from Gregory Sharp, Director, Information and Education, Connecticut Department
of Environmental Protection, 5/25/78.
120. Latham (1975), op. cit., p. 2.
121. At this writing, draft regulations have
been prepared and are being circulated for review.
122. Letter from Jonathan Clapp, Office of Planning and Coordination, Connecticut Department of Environmental Protection, 5/23/78.
123. Extracted from questionnaire information.
124. Tryzna (1974), op. cit., p. 11.
125. Extracted from questionnaire information.
126. Office of Environmental Quality Control
Budget, fiscal 1976-77.
127. Personal communication, Richard O'Connell,
Director, Hawaii Office of Environmental Quality
Control, 2/29/78.
128. Extracted from questionnaire information.
129. Personal communication, Ralph Pickard,
Technical Secretary, Indiana Environmental Management Board, 2/21/78.
130. Letter from Joseph Knapp, Maryland Department of Natural Resources, 5/18/78.
131. Personal communication, James McConnaughay,
State Clearinghouse, Maryland Department of State
Planning, 2/17/78.
132. Yonker, Terry. The Man/Environment Relationship in the Human Ecosystem, paper presented to the
National Association of Environmental Professionals
February 1977, Washington, DC.

133. Letter from Boyd Kinzley, Executive Secretary, Michigan Environmental Review Board, 5/23/78.
134. Personal communication, Terry Yonker, former Executive Secretary, Michigan Environmental Review Board, 3/3/78.
135. Ibid.
136. Letter from Boyd Kinzley, 5/23/78 (see note 133).
137. Perlmutter (1976), op. cit., pp. 51-52.
138. Extracted from questionnaire information.
139. Letter from Duane Noel, Ecological Researcher, Montana Environmental Quality Council, 5/25/78.
140. Extracted from questionnaire information.
141. Ibid.
142. Personal communication, Larry Schmidt, Office of Environmental Review, N.J. Department of Environmental Protection, 5/23/78.
143. Extracted from questionnaire information.
144. Personal communication, Larry Schmidt, 4/12/78 (see note 142).
145. Ibid.
146. Ibid.
147. Extracted from questionnaire information.
148. Extracted from questionnaire information.
149. Personal communication, Harold Lenhart, South Dakota Department of Environmental Protection, 2/14/78.
150. Ibid.
151. Personal communication, Albert Schutz, Planner, Office of the Governor, 3/10/78.
152. Ibid.
153. Ibid.
154. Personal communication, Chauncey Powis, Environmental Coordinator, Office of the Utah State Planning Coordinator, 3/1/78.
155. Ibid.
156. Extracted from questionnaire information.
157. The 1976 Session of the Virginia General Assembly strengthened the Virginia Environmental Quality Act in the area of the Council's responsibility for permit coordination in cases where more than one environmental permit is required for a proposed activity. Interim procedures for implementation of the Multiple Permit Program have been adopted and in effect since January 1978.
158. Tryzna (1974), op. cit., p. 16.
159. Extracted from questionnaire information.
160. Letter from Reginald Wallace, EIS Coordinator, Virginia Council on the Environment, 5/23/78.
161. Commonwealth of Virginia, Joint Environmental Agencies' Budget for the 1978-1980 Biennium.

162. Personal communication, Reginald Wallace, 2/22/78 (see note 160).
163. Ibid.
164. Joint Budget, op. cit. (see note 161).
165. Personal communication, Reginald Wallace, 2/22/78 (see note 162).
166. Letter from Robert Dreis, Bureau of Environmental Impact, Wisconsin Department of Natural Resources, 6/21/78.
167. Ibid.
168. Personal communication, Caryl Terrell, State WEPA Coordinator, Office of State Planning and Energy, 2/15/78.
169. "Implementation of the Wisconsin Environmental Policy Act," prepared by the Interagency WEPA Coordinating Committee,November 1975.
170. Personal communication, Caryl Terrell, 2/15/78 (see note 168).
171. Department of Administration, Department of Agriculture, Department of Business Development, Department of Health and Social Services, Department of Local Affairs and Development, Department of Military Affairs, Educational Communications Board, State Board of Vocational, Technical and Adult Education, Historical Society and Wisconsin Board of Soil and Water Conservation Districts.
172. Department of Employee Trust Funds, Department of Justice, Department of Public Instruction, Department of Regulation and Licensing, Department of Revenue, Department of Veterans Affairs, Commissioner of Banking, Employment Relations Commission, Commissioner of Credit Unions, Higher Education Aids Board, Commissioner of Insurance, Board of Aging, Investment Board, Commissioner of Savings and Loan, Commissioner of Securities, Department of Industry, Labor and Human Relations.
173. Implementation of WEPA (1975), op. cit., p. 6.
174. Letter from Howard Druckenmiller, Director, Bureau of Environmental Impact, Wisconsin Department of Natural Resources, 5/25/78.
175. Letter from Robert Dreis, 6/21/78 (see note 166).
176. Ibid.
177. Personal communication, Caryl Terrell, 2/15/78 (see note 168).
178. Letter from Howard Druckenmiller, 5/25/78 (see note 1974).

Table Sources

3.3   Washington State Department of Ecology.
3.4   California EIR Monitor.
4.1   <u>The California Environmental Quality</u> Act; An
       Evaluation Emphasizing its Impact Upon Cali-
       fornia Cities and Counties with Recommenda-
       tions for Improving its Effectiveness, Volume
       III--Discussion of Costs and Delays Related to
       Environmental Review, prepared by Environment-
       al Analysis Systems, Inc., San Diego, CA,
       November 1975, p. 18.
4.2   CEQA, An Evaluation, p. 7 (see note 15).
4.3   CEQA, An Evaluation, p. 38 (see note 15).
4.4   Personal communication, Norman Hill, Assistant
       Secretary, The California Resources Agency,
       11/18/77.
4.5   Personal communication, Don Meixner, Jake
       Angel, and John McClurg, Department of Water
       Resources, 11/18/77.
4.6   "Implementation of the State Environmental
       Policy Act: A Report to the Legislature,"
       prepared by the Washington (State) Office of
       Program Planning and Fiscal Management,
       November 24, 1976, p. 3.
4.7   Washington State Department of Ecology.
4.8   Personal communication, Dennis Lundblad and
       Tom Elwell, Environmental Review Section,
       Washington Department of Ecology, 11/21/77.
4.9   Personal communication, Bernie Chaplin, Pro-
       ject Planning Supervisor, Washington Depart-
       ment of Transportation, 11/22/77.
4.10 Derived from questionnaire information.
4.11 Personal communication, Thayer Broili and
       Keith Whitenight, North Carolina Department
       of Natural and Economic Resources, 8/76.
4.12 Personal communication, Byron O'Quinn, En-
       vironmental Planning Engineer, North Carolina

Department of Transportation, 8/76.

4.13 Massachusetts Environmental Monitor.

4.14 Massachusetts Environmental Monitor.

4.15 Personal communication, William Hicks, Chief Planner, Massachusetts Office of Environmental Affairs, 12/19/77.

4.16 Personal communication, Richard Bates, Massachusetts Department of Environmental Quality Engineering, 1/26/78.

4.17 Personal communication, John Hurley, Environmental Engineer, Massachusetts Department of Public Works, 12/19/77.

4.18 Personal communication, David Russell and Glen Yee, Massachusetts Department of Public Utilities, 8/76.

4.19 "Environmental Impact Statement Preparation: Is a Central Office the Answer?" A report to the Legislature by the Minnesota State Planning Agency/Environmental Quality Board, February 15, 1978.

4.20 Personal communication, Charles Kenow and Nancy Onkka, Minnesota Environmental Quality Council, 8/76.

4.21 Personal communication, Dale McMichael, Coordinator, Office of Environmental Analysis, Minnesota Pollution Control Agency, 8/76.

4.22 Personal communication, Pat Grady, N.Y.S. Department of Environmental Conservation, 12/8/77.

4.23 Budget information and personal communication, Pat Grady (see note 36).

4.24 Personal communication, Pat Grady (see note 36).

5.1 Michigan Environmental Review Board.

5.2 Extracted from questionnaire information.

5.3 Personal communication, Duane Noel, Montana Environmental Quality Council, 5/25/78.

5.4 Personal communication, Larry Schmidt, Office of Environmental Review, New Jersey Department of Environmental Protection, 5/23/78.

5.5 Extracted from questionnaire information.

5.6 "Implementation of the Wisconsin Environmental Policy Act," prepared by the Interagency WEPA Coordinating Committee, November, 1975.

Figure Sources

4.1   "Regulations Governing the Implementation of the Massachusetts Environmental Policy Act." May 16, 1978.

4.2   "Environmental Impact Statement Preparation: Is a Central Office the Answer?" A Report to the Legislature by the Minnesota State Planninς Agency/Environmental Quality Board, February 15, 1978.

4.3   Guidelines, Environmental Impact Assessment and Statements, Office of Parks and Recreation, 1/77.

5.1   Hawaii Environmental Quality Council.

5.2   Michigan Environmental Review Board.

5.3   "The Environment: Policy--Guidelines and Procedures for Processing Environmental Impact Statements," November 1975, p. 26.

5.4   "Procedures Manual and Guidelines for the Environmental Impact Statement Program in the Commonwealth of Virginia," June 1976, p. 6.

5.5   Joint Environmental Agencies' Budget for the 1978-1980 Biennium.

5.6   Wisconsin Environmental Policy Act, Revised Guidelines, February 1976.

References

Anderson, Frederick. NEPA in the Courts, Baltimore:
    The Johns Hopkins University Press, 1973.
Andrews, Richard, Paul Cromwell, Gordon Enk, Edward
    Farnworth, James Hibbs, and Virginia Sharp.
    Substantive Guidance for Environmental Impact
    Assessment: An Exploratory Study, The Insti-
    tute of Ecology, 1977.
Burchell, R. and Listokin, D. The Environmental Im-
    pact Handbook, Center for Urban Policy Research,
    Rutgers--The State University, New Brunswick,
    NJ, 1975.
Dreyfus, Daniel and Helen Ingram. "The National En-
    vironmental Policy Act: A View of Intent and
    Practice," Natural Resources Journal, Vol. 16,
    No. 2, April 1976.
Duvernoy, Eugene. "A Guide to the New York State
    Environmental Quality Review Act," Center for
    Environmental Research, Cornell University,
    Ithaca, NY, March 1978.
Enk, Gordon A. Beyond NEPA: Criteria for Environ-
    mental Impact Review, The Institute on Man and
    Science, Rensselaerville, NY, May 1973.
Enk, Gordon A. and Stuart Hart. Beyond NEPA Re-
    visited: Directions in Environmental Impact
    Review, The Institute on Man and Science,
    Rensselaerville, NY, March 1978.
Environmental Impact Statements, A Report of the Com-
    mission on Federal Paperwork, Washington, DC,
    February 25, 1977.
Environmental Impact Statements: An Analysis of Six
    Years' Experience by Seventy Federal Agencies,
    Report of the Council on Environmental Quality,
    March 1976.
"Environmental Impact Statement Preparation: Is a
    Central Office the Answer?" A report to the
    Legislature by the Minnesota State Planning

Agency/Environmental Quality Board, February
15, 1978.
Environmental Quality: The First Annual Report of
    the Council on Environmental Quality, Washing-
    ton, DC: GPO, August 1970.
Environmental Quality: The Second Annual Report of
    the Council on Environmental Quality, Washing-
    ton, DC: GPO, August 1971.
Environmental Quality: The Third Annual Report of
    the Council on Environmental Quality, Washing-
    ton, DC: GPO, August 1972.
Environmental Quality: The Fourth Annual Report of
    the Council on Environmental Quality, Washing-
    ton, DC: GPO, August 1973.
Environmental Quality: The Fifth Annual Report of
    the Council on Environmental Quality, Washing-
    ton, DC: GPO, August 1974.
Environmental Quality: The Sixth Annual Report of
    the Council on Environmental Quality, Washing-
    ton, DC: GPO, August 1975.
Environmental Quality: The Seventh Annual Report of
    the Council on Environmental Quality, Washing-
    ton, DC: GPO, August 1976.
Green, Harold P. The National Environmental Policy
    Act in the Courts: January 1, 1970-April 1,
    1972, Washington, DC: The Conservation Founda-
    tion, 1972.
Guidelines for Environmental Impact Assessment, Ter-
    restrial Resources, State University of New
    York College of Environmental Science and
    Forestry, Syracuse, NY, August 1975.
Heffernan, P. and R. Corwin (eds.). Environmental
    Impact Assessment, Freeman, Cooper & Co.,
    San Francisco, CA, 1975.
"Implementation of the Minnesota Environmental Poli-
    cy Act," prepared for the Minnesota Environ-
    mental Quality Council by Hayworth and Ander-
    son, Inc., 1974.
"Implementation of the State Environmental Policy
    Act: A Report to the Legislature," prepared
    by the Washington (State) Office of Program
    Planning and Fiscal Management, November 24,
    1976.
"Implementation of the Wisconsin Environmental Poli-
    cy Act," prepared by the Interagency WEPA Co-
    ordinating Committee, November 1975.
Improving the Environmental Impact Assessment Process,
    University of California, Berkeley, College of
    Environmental Design, December 1972.
Latham, Janis. "Connecticut Environmental Policy
    Act (CEPA): Problems with Implementation;

Proposals for Revision," Office of Legislative Research, October 1975.

Leopold, Luna, Frank Clarke, Bruce Hanshaw and James Balsley. A Procedure for Evaluating Environmental Impact, U.S. Geological Survey Circular 645, Washington, DC: U.S. Geological Survey, 1971.

Liroff, Richard. A National Policy for the Environment: NEPA and its Aftermath, Indiana University Press, Bloomington, 1976.

McHarg, Ian L. Design with Nature, Garden City, NY: Doubleday & Co., Inc. (Doubleday Natural History Press), 1969.

Natural Resources Defense Council. Land-Use Controls in the United States, New York, NY: The Dial Press, 1977.

Orloff, Neil. The Environmental Impact Statement Process, Office of Federal Activities, U.S. EPA, February 1973.

Pearlman, Kenneth. "Little NEPA's: The Impact Statement at the State Level," Department of City and Regional Planning, The Ohio State University. Prepared for presentation to the American Institute of Planners, 1975.

Perlmutter, Steven. The Montana Environmental Policy Act: The First Five Years, Montana State Legislature/Environmental Quality Council, Helena, MT, November 29, 1976.

Richardson, Dan. The Cost of Environmental Protection, Center for Urban Policy Research, Rutgers--The State University, New Brunswick, NJ, 1976.

Roe, Charles. "The North Carolina Environmental Policy Act: Neglected Planning Tool," paper submitted to the Institute of Government, Chapel Hill, NC, April 1975.

Smarden, R., J.R. Pease and P. Donheffner. Environmental Assessment Resource Handbook, Oregon State University Extension Service, Corvallis, OR, September 1976.

_____. Environmental Assessment Manual, Special Report 465, Oregon State University Extension Service, Corvallis, OR, September 1976.

"State Environmental Quality Review Act: Handbook for Local Government," prepared by the New York State Department of Environmental Conservation, October 1976.

Steck, Henry. NEPA After Five Years: A Second Look, Department of Political Science, SUNY-Cortland, prepared for Annual Meeting of the Society for the Study of Social Problems, 1975.

249

The California Environmental Quality Act: A Review,
prepared by the State Office of Planning and
Research, Sacramento, CA, March 1976.

The California Environmental Quality Act; An Evaluation Emphasizing its Impact Upon California
Cities and Counties with Recommendations for
Improving its Effectiveness, Volume III--Dis-
cussion of Costs and Delays Related to Environmental Review, prepared by Environmental Analysis Systems, Inc., San Diego, CA, November 1975.

"The California Environmental Quality Act: The Integration of CEQA and Local Planning Practices,"
prepared by the Public Policy Research Organization, University of California at Irvine,
January 1976.

The Environmental Impact Statement--It Seldom Causes
Long Project Delays But Could Be More Useful
If Prepared Earlier, a Report to the Committee
on Environment and Public Works by the U.S.
General Accounting Office, August 9, 1977.

Tryzna, Thaddeus. Environmental Impact Requirements
in the States: NEPA's Offspring, prepared for
the Office of Research and Development, U.S.
EPA, April 1974.

Tryzna, Thaddeus and Arthur Jokela. California Environmental Quality Act: Innovation in State
and Local Decisionmaking, prepared for the
Office of Research and Development, U.S. EPA,
October 1974.

Yost, Nicholas. "NEPA's Progeny: State Environmental Policy Acts," Environmental Law Reporter,
Vol. 3, 1973, pp. 50090-50098.

Appendix A
SEPA Fact Sheets

CALIFORNIA

*California Environmental Quality Act of 1970, Cal.
Pub. Res. Code, Section 21000-21176 (Supp. 1972),
as amended by: Ch. 1154, Statutes of 1972, December
5, 1972; Ch. 895, Statutes of 1973, September 28,
1973; Ch. 56, Statutes of 1974, March 4, 1974; Ch.
176, Statutes of 1974, May 21, 1974; Ch. 1187, Statutes of 1975, September 30, 1975; Ch. 593, Statutes
of 1976, August 27, 1976; Ch. 753, Statutes of 1976,
September 7, 1976; Ch. 1312, Statutes of 1976, September 29, 1976.*

> State Contact: Norman E. Hill, Assistant
> to the Secretary for Resources, The Resources
> Agency, 1414 Ninth Street, Sacramento, CA,
> 95815 (Phone: 916-445-9134).

LEGISLATION

1. Extensiveness of Coverage: Impact statements are
    required for any action "carried out or approved
    by public agencies." This has come to be inter-
    preted as including not only actions undertaken
    or funded by state and local agencies, but also
    private projects for which state or local govern-
    mental permission is required.
2. Criteria for EIS Preparation: Only those actions
    deemed as having the potential for "significant
    effect on the physical environment" are required
    to comply with the action forcing provisions.
3. Departure from or Addition to the Language of
    NEPA: There are significant additions to NEPA
    language (see Noteworthy Features). In terms of
    EIS content, the following entries supplement the
    traditional five NEPA considerations:

    (a) Mitigative measures proposed to minimize the
        significant environmental effects including,
        but not limited to, measures to reduce waste-
        ful, inefficient, and unnecessary consumption
        of energy.
    (b) The growth-inducing impact of the proposed
        project.

    The report shall also contain a statement briefly
    indicating the reasons for determining that vari-
    ous effects of a project are not significant and
    consequently have not been discussed in detail in
    the environmental impact report.

REGULATION

1.1. Guidelines or Rules and Regulations: Most re-
     cently, "State EIR guidelines," Cal. Administra-
     tive Code, Title 14; issued December of 1976 by
     the Resources Agency.
1.2. Interpretation of Legislation:
     A. EIS Content: The guidelines expand upon the
        requirements set forth in the legislation.
        In addition, however, they do require that
        "EIR's include a discussion of the potential
        energy impacts of proposed projects, with
        particular emphasis on avoiding or reducing
        inefficient, wasteful and unnecessary con-
        sumption of energy."
     B. Threshold Criteria: The term "significant
        effect" has been defined as a substantial or
        potentially substantial adverse change in
        the environment. Appendix G of the State
        EIR Guidelines contains a list of examples
        of activities which would normally be found
        to be significant. Consideration of long-
        term effects, cumulative impacts, and indi-
        rect consequences are stressed.
     C. Exemptions: The guidelines provide categori-
        cal exemptions for twenty classes of actions
        determined not to have a significant effect.
        These include such actions as repair or re-
        placement of existing facilities, minor al-
        terations to land and other actions such as
        environmentally protective regulatory activi-
        ties. In addition to these, blanket exemp-
        tions are also provided for emergency actions
        and ministerial projects.

ADMINISTRATION

1.1. Central Coordinating Agency: The Resources Agen-
     cy is responsible for the administration of the
     Act. The Office of Planning and Research, how-
     ever, in addition to acting as the Clearinghouse,
     occasionally gets involved in review and is cur-
     rently rewriting the CEQA handbook.
1.2. Does the Coordinating Agency Possess the power...
     A. to establish guidelines or rules and regula-
        tions? Yes, pursuant to CEQA.
     B. to review agency implementing regulations?
        No.
     C. to make threshold determinations? No. Pro-
        ject originating or lead agencies determine
        need for EIR according to their own criteria.

D. to reject inadequate EIS's? No. The guidelines do require, however, that whenever a
state agency plans to issue a permit for a
local agency project which is subject to an
EIR, the EIR must be sent to the State for
review.
2. Coordination of EIS Requirement with Permit
Process: Yes. The program is designed to produce a final EIR which will be presented to
the decision-making body of the Agency when it
decides to carry out a project or issue a permit.
3. Coordination of SEPA with NEPA: All or any part
of an EIS prepared under NEPA may be submitted
in lieu of all or any part of an EIR prepared
under CEQA, so long as the material complies
with State regulations.

PARTICIPATION

1. Public Solicitation and Involvement:
A. in the development of guidelines or rules
and regulations: The EIR Guidelines were
developed and subsequently amended through
an open hearing process. The oral and written comments resulting from these hearings
led to many alterations.
B. in the review of environmental documents:
While CEQA does not require formal public
hearings at any stage of the environmental
review process, notice of the preparation of
EIR's and negative declarations is required
to be posted and published in a newspaper
of general circulation. The Resources
Agency also publishes a weekly California
EIR Monitor to provide notice of amendments
to the guidelines, the completion of draft
EIR's, and other information as deemed appropriate.

LITIGATION

Extensive; some 244 lawsuits were filed between 1970
and 1975. The lawsuits involved the same range of
issues as have been addressed in NEPA litigation.
The cases involved both public and private actions
and both procedural and substantive issues; nearly
all decisions, however, have been based on procedural issues.

Probably of greatest significance has been the <u>Friends of Mammoth vs. Board of Supervisors of Mono County</u> decision of 1972 which, like NEPA, held the law applicable not only to governmental actions, but also to the approval by government of potentially significant private activities.

<u>NOTEWORTHY FEATURES</u>

- CEQA is not limited to "major actions." Minor actions may be subject to the act if they might have a significant effect on the environment. For example, a small governmental activity which might eliminate a rare or endangered species would be subject to the requirement for an EIR.
- The law states that projects should not be approved if there are "feasible alternatives or feasible mitigation measures available which would substantially lessen the significant environmental effects." However, in the event "specific economic, social, or other conditions make infeasible such project alternatives or such mitigation measures, individual projects may be approved in spite of one or more significant effects thereof."
- Two sections (21002.1 and 21100.1) from the 1976 amendments to the law greatly focus the purpose and scope of environmental impact reports prepared under CEQA (see Text).
- CEQA establishes an explicit statute of limitation concerning commencement of legal actions or proceedings.
- A unique section (21160.) dealing with the submission of data and information by private applicants provides that any such information definable as a "trade secret" under Section 6254.7 of the government Code shall not be included in the impact report or otherwise disclosed.
- California is unique in that a number of studies have been prepared on various aspects of CEQA (see References). Most other states have not, as yet, initiated such an effort of documentation.

256

CONNECTICUT

Connecticut Environmental Policy Act of 1973, Public Act. 73-562, Conn. Gen. Stat. Ann. Ch. 439, Section 22a-1, et seq. (Cum. Supp. 1974-75), as amended 1975 and 1977.

State Contact: Gregory Sharp. Director, Information and Education, Department of Environmental Protection, State Office Building, Rm. 114, Hartford, CN 06115 (Phone: 203-566-5524)

LEGISLATION

1. Extensiveness of Coverage: The EIS prerequisite applies only to those projects directly undertaken by "state departments, institutions, or agencies, or funded in whole or in part by the state."

2. Criteria for EIS Preparation: EIS's are to be prepared only on "actions which may significantly affect the environment." Significant actions are further defined as actions "which could have a major impact on the state's land, water, air or other environmental resources, or could serve short term to the disadvantage of long term environmental goals."

3. Departure from or Addition to the Language of NEPA: CEPA significantly augments the traditional five EIS requirements of NEPA by inserting entries for the consideration of:
    (a) direct and indirect effects which might result during and subsequent to the proposed action.
    (b) mitigation measures proposed to minimize environmental impacts.
    (c) an analysis of the short term and long term economic, social, and environmental costs and benefits of the proposed action.
    (d) the effect of the proposed action on the use and conservation of energy resources.

    CEPA also bolsters the traditional consideration of "alternatives to the proposed action" by requiring the consideration of the "no-build" alternative.

REGULATION

1.1. Guidelines or Rules and Regulations: Regulations have been prepared (May 1978) by the Department of Environmental Protection (DEP) pursuant to the amendments of 1977 and are being circulated for review and comment.

1.2. Interpretation of Legislation:
   A. EIS Content: DEP informally assumes the responsibility for developing policy and procedure until regulations are issued.
   B. Threshold Criteria: Pending the issuance of the regulations. Until such regulations are in place, the potential for litigation continues to be the primary concern of administrators. Thus, the degree of controversy is currently most often used to determine need for an EIS.
   C. Exemptions: Pending the issuance of the regulations. Amendments to the law in 1977 did, however, categorically exempt emergency measures and actions of a ministerial nature.

ADMINISTRATION

1.1. Central Coordinating Agency: Although the legislation authorized the Council on Environmental Quality (similar to that at the Federal level) to review and make recommendations on EIS's, no provision for staff was made. Therefore, the bulk of EIS review and coordination takes place within DEP.

1.2. Does the Coordinating Agency Possess the power...
   A. to establish guidelines or rules and regulations? Yes, pursuant to the amendments of 1977.
   B. to review agency implementing regulations? No.
   C. to make threshold determinations? Informally, pending codification by the forthcoming regulations.
   D. to reject inadequate EIS's? Although coordination of the EIS review process takes place within DEP, it does not possess enforcement powers. The final determination as to whether project evaluation satisfies the requirements of CEPA is made by the Office of Policy and Management, an appendage to the Governor's Office.

2. Coordination of EIS Requirement with Permit Process: A lack of clarity in the procedures for

EIS implementation currently inhibits such co-ordination. Neither the law nor the requirements for the regulations adequately address this issue.
3. Coordination of SEPA with NEPA: The law specifically exempts any projects for which EIS's have previously been prepared under other state or Federal laws or regulations.

PARTICIPATION

1. Public Solicitation and Involvement:
   A. in the development of guidelines or rules and regulations: None.
   B. in the review of environmental documents: The review process detailed by the Act requires the submission of EIS evaluations to affected municipalities for publication of availability. Evaluations are also to be published in the Connecticut Law Journal. A public hearing is required if twenty-five persons request such a hearing within ten days of publication.

LITIGATION

None.

NOTEWORTHY FEATURES

- The 1977 amendments to CEPA require that a public hearing be held on an EIS if "twenty-five persons or an association having not less than twenty-five persons request such a hearing within ten days of the publication of the notice in the Connecticut Law Journal."

HAWAII

State Environmental Policy Act of 1974 (Hawaii Rev. Stat. Ch. 344); and Environmental Quality Commission and Environmental Impact Statements Act of 1974 (Hawaii Rev. Stat. Ch. 343)

State Contact: Richard O'Connell, Director, Office of Environmental Quality Control, Office of the Governor, 550 Halekauwila St., Rm. 301, Honolulu, HI 96813 (Phone: 808-548-6915)

LEGISLATION

1. Extensiveness of Coverage: The EIS requirement applies to actions involving "the use of state or county lands or the use of state or county funds," as well as to a limited range of private activities, namely:
    (a) any activity in the Waikiki area,
    (b) any amendment to state or county general plans(zoning),
    (c) any activity within a State-designated Conservation District,
    (d) any historic site,
    (e) all shorline areas.
2. Criteria for EIS Preparation: Consideration of "environmental factors and available alternatives" is to be incorporated in the "feasibility or planning studies" for any action "which will probably have significant effects."
3. Departure from or Addition to the Language of NEPA: The law defers to the rules and regulations for specifying EIS content. The review mechanism is, however, more clearly defined than under NEPA (see "Administration" below).

REGULATION

1.1. Guidelines or Rules and Regulations: "Rules of Practice and Procedure" and "Environmental Impact Statement Regulations," promulgated by the Hawaii Environmental Quality Commission (EQC) in 1975.
1.2. Interpretation of Legislation:
    A. EIS Content: The rules and regulations establish the contents of EIS's. Several additional entries are included:
        (a) the relationship of the proposed action

to land-use plans, policies, and controls of the affected area.
(b) an indication of what other interests
and considerations of governmental
policies are thought to offset the adverse environmental effects of the proposed action.
(c) organizations and persons consulted.
B. Threshold Criteria: A "significant effect"
may vary with individual setting and circumstances of particular actions. Generally,
however, any action which may have a major
effect on the quality of the environment, or
affect the economic or social welfare of an
area, or would possibly be contrary to the
State's environmental policies or long-term
environmental goals and guidelines as expressed in Chapter 342 and 344, Hawaii Revised Statutes, and any revisions thereof
and amendments thereto, would likely result
in a "significant effect."
C. Exemptions: The regulations specify actions
exempt from the EIS requirement such as re-
pairs and maintenance. They do not include
blanket exemptions for ministerial or emergency actions. Agencies are encouraged to
develop their own specific list of exempt
actions.

ADMINISTRATION

1.1. Central Coordinating Agency: The legislation
established an Environmental Quality Commission
(EQC) to develop implementing rules and regulations, and administer the law. The members of
the Commission serve without pay and are supported by a two-person staff which is provided
by the Office of Environmental Quality Control
(OEQC). OEQC reviews and comments on all EIS's
in the State and serves as informal Clearinghouse for Federal EIS's.
1.2. Does the Coordinating Agency Possess the power...
A. to establish guidelines or rules and regulations? Yes, pursuant to Ch. 343.
B. to review agency implementing regulations?
EQC Regulations are binding on all agencies.
The Commission does, however, review individual agency proposals for exempt actions.
C. to make threshold determination? No. Proposing agencies determine need for EIS (or
the agency approving the action in the case

of a private applicant).
D. to reject inadequate EIS's? The Commission provides recommendations as to the acceptability of impact statements. The final authority to accept or reject rests with:
  (1) the governor, or his authorized representative, whenever an action proposes the use of state lands or the use of state funds; or
  (2) the mayor, or his authorized representative, of the respective county whenever an action proposes only the use of county lands or county funds.
2. Coordination of EIS Requirement with Permit Process: Permits are not issued on activities requiring an EIS. The regulations do, however, require that the EIS process be coordinated with the "land use plan, policies, and controls of the affected area."
3. Coordination of SEPA with NEPA: Whenever an action is subject to both NEPA and the state EIS requirement, the draft statement must be submitted to EQC for distribution and review thirty days prior to submission to the President's Council on Environmental Quality. Likewise, the final statement must first be approved by the governor or the applicable mayor before the statement is forwarded to CEQ.

PARTICIPATION

1. Public Solicitation and Involvement:
A. in the development of guidelines or rules and regulations: Four months of public hearings on the draft regulations provided opportunities for persons on each Island to attend a hearing at a distance no greater than 45 miles from home.
B. in the review of environmental documents: Public comment is solicited by distributing EIS's to community associations within the project area and by publishing the availability of EIS's in a bi-weekly newsletter, "EQC Bulletin," sent to some 500 agencies, organizations, or individuals. A thirty day comment period is established for EIS's and any person may request to be consulted during the actual drafting of the EIS.

LITIGATION

Negligible; most cases have focused on procedural issues.

NOTEWORTHY FEATURES

- Hawaii's statutes separate the statement of environmental policy (Chapter 344) from the requirement for EIS's (Chapter 343).
- Chapter 343 encourages the incorporation by reference of previous determinations and impact statements.
- Chapter 343 imposes a statute of limitations on any judicial proceedings resulting from the EIS process.
- Chapter 343 mandates that "at least one public hearing shall be held in each county" prior to the final adoption, amendment, or repeal of implementing rules and regulations.
- The policy statement includes a unique provision for the setting of population limits.

INDIANA

Indiana Environmental Policy Act of 1972, Pub. L. 98, 1972, Ind. Stat. Ann. Section 35-5301 et. seq. (Supp. 1971)

> State Contact: Ralph Pickard, Technical Sec-
> retary, Environmental Management Board, 1330
> W. Michigan Street, Indianapolis, IN 46206
> (Phone: 317-633-8404) or L. Robert Carter, Co-
> ordinator of Environmental Programs, same ad-
> dress as above (Phone: 317-633-8467).

LEGISLATION

1. Extensiveness of Coverage: The EIS requirement applies only to actions directly undertaken or funded by state agencies.
2. Criteria for EIS Preparation: Impact statement preparation is required only for actions "significantly affecting the quality of the human environment."
3. Departure from or Addition to the Language of of NEPA: None.

REGULATION

1.1. Guidelines or Rules and Regulations: "EMB-2: The Definition of Actions of State Agencies which have Significant Environmental Impact" (effective August 1975) and "EMB-3: Environmental Quality Review" (effective January 1977), both promulgated by the Indiana Environmental Management Board (EMB).
1.2. Interpretation of Legislation:
   A. EIS Content: EMB-3 adds the following considerations to EIS content:
      (a) Growth Inducement Aspects
      (b) Measures Proposed to Mitigate Adverse Effects of the Action
      (c) Effect on the Use and Conservation of Energy Resources
   B. Threshold Criteria: EMB-2 establishes a preliminary assessment procedure for determining which actions significantly affect the environment. Consideration of "both primary and secondary consequences of short-term and long-term duration" are stressed as are cumulative impacts. The potential for

controversy is also considered justification for EIS preparation.

C. Exemptions: Categorical exemptions developed under EMB-2 include minor actions and emergency actions. Agencies are encouraged to submit their own list of exemptions for approval by EMB.

ADMINISTRATION

1.1. Central Coordinating Agency: The Environmental Management Board (EMB) is responsible for the coordination of the program, including the maintenance of rules and regulations. The eleven unpaid members of the Board are supported by a single staff member assigned to the Technical Secretary of the Agency.

1.2. Does the Coordinating Agency Possess the power...

A. to establish guidelines or rules and regulations? Yes, pursuant to the legislation which created it (Environmental Management Act of 1971).

B. to review agency implementing regulations? Yes.

C. to make threshold determinations? No. Project originating agencies determine the need for EIS according to EMB-2.

D. to reject inadequate EIS's? Yes. After other agencies and the public have commented on specific issues or areas of expertise during the draft process, EMB reviews the final EIS to ensure that specified procedures were followed and that comments received in the draft stage were properly addressed. A decision whether to accept the final EIS or require it to be revised is then made on this basis. Thus, EMB has the power to reject EIS's but depends on the substantive comments of other parties.

2. Coordination of EIS Requirement with Permit Process: Projects requiring permits or licenses are statutorily exempt from EIS requirements.

3. Coordination of SEPA with NEPA: Any action requiring an EIS under NEPA is exempt from the Indiana law unless the action requires State legislation or State appropriations.

PARTICIPATION

1. Public Solicitation and Involvement:

A. in the development of guidelines or rules

and regulations: A public hearing was held on the draft regulations.

B. in the review of environmental documents: Draft EIS's are to be circulated to local, state, and Federal agencies and to the general public as deemed necessary by the project originating agency. On the basis of comments received, the agency then determines whether or not to conduct a public hearing on the EIS.

LITIGATION

Negligible; one case (J.M. Foster Co., Inc. vs. Northern Indiana Public Service Co., Inc.) however, tested the applicability of the law. Complaintant alleged that the electric utility was a "state agency" because of right of eminent domain and should therefore be required to prepare EIS's.

NOTEWORTHY FEATURES

- The Act legislatively exempts any action involving "the issuance of a license or permit by any agency of the state."

SEPA FACT SHEET

MARYLAND

Maryland Environmental Policy Act of 1973, Ch. 702, Md. Acts of 1973, 41 Ann. Code of Md., Section 447-451, (Cum. Supp. 1973), and Ch. 703, Md. Acts of 1973 Natural Res. Art., Ann. Code of Md., Section 1-301 et seq. (1974 Volume) as amended by Ch. 129 of the Md. Acts of 1975, Section 1-301(c)

State Contact: Joseph Knapp, Administrator, Clearinghouse Review, Department of Natural Resources, Tawes State Office Building, Annapolis, MD 21401 (301-269-3548).

LEGISLATION

1. Extensiveness of Coverage: The EIS requirement applies only to proposed state actions such as appropriations of state funds and legislative proposals.
2. Criteria for EIS Preparation: Proposed state actions "significantly affecting the quality of the environment" require the preparation of an "environmental effects report"(EER).
3. Departure from or Addition to the Language of NEPA: MEPA uniquely defines its EER as a "report on each proposed State action significantly affecting the environment, natural as well as socioeconomic and historic." The definition of impact statement content also greatly differs from the traditional NEPA model: EER's are to include, but not be limited to, discussions of beneficial as well as adverse environmental effects, the goal being maximize the former and minimize the latter.

REGULATION

1.1. Guidelines or Rules and Regulations: "Revised Guidelines for Implementation of the Maryland Environmental Policy Act" issued by the Secretary of the Department of Natural Resources (DNR) in June of 1974.
1.2. Interpretation of Legislation:
  A. EIS Content: Although the law implies it, the guidelines specify that an EER is not required for the issuance of individual licenses or permits. Consideration of beneficial effects and cumulative significance

267

is also stressed in the preparation of
EER's.

    B. Threshold Criteria: A preliminary environ-
mental assessment form (EAF) developed by
DNR is used by departmental units to determine the "significance" of their "proposed actions." If the impact is determined
to be significant, either adverse or beneficial, an EER is required.

    C. Exemptions: There is no provision of "cate-
gorical exemption."

ADMINISTRATION

1.1. Central Coordinating Agency: The legislation
designates the Department of Natural Resources
(DNR) as responsible for the promulgation of
general implementing guidelines. The State
Clearinghouse in the Department of State Planning (DSP), however, is responsible for coordinating the review of EER's prepared under
MEPA. Thus, coordinating responsibilities are
split.

1.2. Does the Coordinating Agency Possess the power...

    A. to establish guidelines or rules and regula-
tions? Yes. DNR fills this role.

    B. to review agency implementing regulations?
No. Agency regs must, however, be consistent with the guidelines promulgated by DNR.

    C. to make threshold determinations? No. Pro-
posing agencies determine need for EIS
through EAF form developed by DNR.

    D. to reject inadequate EIS's? No. Although
DSP conducts its own review of EER's so as
to determine the adequacy of information presented, it can only make recommendations
based on that review.

  2. Coordination of EIS Requirement with Permit Process: The EER requirement does not extend to
actions involving individual licenses or permits.

  3. Coordination of SEPA with NEPA: An EIS prepared under NEPA may be submitted as an EER as
long as it contains the information required by
the guidelines.

PARTICIPATION

1. Public Solicitation and Involvement:

    A. in the development of guidelines or rules and
regulations: None.

    B. in the review of environmental documents:

Environmental documents are available at
cost from the preparing agency; the State
Clearinghouse prepares a quarterly list of
all such documents and distributes it to:
(a) all members of the State General Assembly;
(b) all State departments;
(c) all newspapers, private citizens and
    citizen groups who request such listing.
A public information meeting may be held
through formal written request from a local
government or from fifty or more citizens.

LITIGATION

Negligible; a single case in 1976, however, affirmed
the fact that the law does not apply to private activities.

NOTEWORTHY FEATURES

- MEPA covers only "proposed state actions."
  Its purpose is to provide the State legislature information on environmental effects
  consequent to appropriations of State funds
  and legislative proposals.
- "Significant effects" include impacts on
  the "natural as well as socioeconomic and
  historic" environment.
- Consideration of adverse and beneficial environmental effects is required, the goal
  being to minimize the former and maximize
  the latter.
- A public information meeting may be held
  through formal written request from a local government or from fifty or more citizens concerning an environmental effects report (EER).

SEPA FACT SHEET

MASSACHUSETTS

Ch. 781, Acts of 1972, Ann. Laws Mass., Ch. 30, Sec. 61-62. (Cum. Supp. 1973), as amended by Ch. 257 of the Acts of 1974 and 1978.

State Contact: William Hicks, Director, MEPA Unit, Executive Office of Environmental Affairs, 100 Cambridge Street, Room 2000, Boston, MA 02202 (Phone: 617-727-5830)

LEGISLATION

1. Extensiveness of Coverage: MEPA covers projects or activities initiated, financed, or permitted by state agencies including the actions of local redevelopment authorities, housing authorities, and development commissions.
2. Criteria for EIS Preparation: Only actions deemed as having the potential for causing "damage to the environment" require the preparation of environmental impact reports (EIR's).
3. Departure from or Addition to the Language of NEPA: There are significant additions to NEPA language (see Noteworthy Features). In terms of EIR content, the consideration of "all measures being utilized to minimize environmental damage," supplements the traditional five NEPA considerations.

REGULATION

1.1. Guidelines or Rules and Regulations: "Rules and Regulations for the Preparation of Environmental Impact Reports," dated 1973, as revised 1975 and 1976 by the Executive Office of Environmental Affairs. (EOEA drafted new rules and regulations responsive to the 1978 amendments to the law entitled, "Regulations Governing Implementation of the Massachusetts Environmental Policy Act" in May of 1978).
1.2. Interpretation of Legislation:
   A. EIS Content: The regulations expand upon the requirements set forth in the legislation. The 1978 amendments to the law, however, empower the Secretary of Environmental Affairs to determine both the scope and content of EIR's (see Noteworthy Features).
   B. Threshold Criteria: "Damage to the environ-

ment" is defined in detail (see text).

C. Exemptions: The new rules and regulations required the Secretary of Environmental Affairs to establish categories of projects and permits that will always and will never require impact reports by July 1, 1978.

ADMINISTRATION

1.1. Central Coordinating Agency: The Environmental Impact Review Division, within the Secretary's Office of the Executive Office of Environmental Affairs (EOEA) is responsible for the coordination, implementation, and monitoring of the MEPA process. This includes the responsibility for promulgating rules and regulations.

1.2. Does the Coordinating Agency Possess the power...
   A. to establish guidelines or rules and regulations? Yes, pursuant to MEPA.
   B. to review agency implementing regulations? The 1978 amendments to the law empowered the Secretary to issue rules and regulations binding on all state agencies.
   C. to make threshold determinations? Yes. All project notifications and assessments must be submitted to the Secretary for determination of the need for impact report preparation.
   D. to reject inadequate EIS's? No, although the Secretary is required to comment on each action by law, only the Attorney General's Office can stop a project until the law's obligations are met.

2. Coordination of EIS Requirement with Permit Process: Yes. The preparation of EIR's during the initial planning and design phase is encouraged and permits must be acted upon with ninety days of completion of the MEPA process.

3. Coordination of SEPA with NEPA: Yes. EIS's prepared under NEPA may be submitted as EIR's under MEPA but must comply with the State regulations for environmental review. A negative declaration prepared under NEPA, however, does not constitute compliance.

PARTICIPATION

1. Public Solicitation and Involvement:
   A. in the development of guidelines or rules and regulations: Public hearings are incorporated into the rule-making procedure.

B. in the review of environmental documents:
Aside from the provisions concerning public
notice (through a bi-monthly Environmental
Monitor and newspaper notice requirements)
and receipt of public comment, there are no
formal mechanisms for citizen involvement.
Although public hearings are "encouraged,"
the actual decision as to whether or not a
public hearing should be held is left to the
discretion of the responsible agency.

LITIGATION

Negligible.

NOTEWORTHY FEATURES

- Amendments to the law in 1978 authorized
  the Secretary's Office to issue regulations
  for MEPA binding on all agencies.
- The power of threshold determination resides
  with the Secretary (and hence the Environmental Impact Review Division)--all preliminary assessment documents must be submitted to this centralized body for determination of whether or not an EIR is required.
- The 1978 MEPA amendments empower the secretary, in cooperation with the proposing
  agency or person, to limit the scope of the
  EIR to those issues "which by the nature and
  location of the project are likely to cause
  damage to the environment." Specifically,
  this means that the Secretary shall determine the form, content, level of detail, and
  alternatives required for each report.
- MEPA provides an innovative statute of limitation regarding the commencement of legal
  proceedings; a notice of intention to commence such proceedings is required within
  sixty days of project clearance. This, it
  is hoped, will allow for a period of out-of-
  court problem solving before legal proceedings actually begin.
- One section of the 1978 amendments directs
  the Secretary of Environmental Affairs to
  seek from the Federal Government, the delegation of authority to carry out the National Environmental Policy Act.

272

MICHIGAN

*Michigan Executive Directive 1971-10, as superceded
by Michigan Executive Order 1973-9 as superceded by
Michigan Executive Order 1974-4 (May 1974).*

State Contact: Boyd Kinzley, Executive Secretary, Environmental Review Board, Department of Management and Budget, Lansing, MI
48913 (Phone: 517-373-6491)

EXECUTIVE ACTION

1. Extensiveness of Coverage: EIS's are to be prepared by state agencies proposing "major actions" within their jurisdictions. This is interpreted to include "any policy, administrative
   action, or project;" the term "administrative
   action" includes the issuance of permits and
   utilization of state funds.
2. Criteria for EIS Preparation: Actions that "may
   have a significant impact on the environment or
   human life" are required to have a statement
   prepared.
3. Departure from or Addition to the Language of
   NEPA: The contents of EIS's required under Executive Order 1974-4 differ somewhat from the
   traditional five NEPA requirements:

   Each statement shall contain the following:
   1. A description of the probable impact of
      the action on the environment, including
      any associated impacts on human life.
   2. A description of the probable adverse
      effects of the action which cannot be
      avoided (such as air or water pollution,
      threats to human health or other adverse
      effects on human life).
   3. Evaluation of alternatives to the pro-
      posed action that might avoid some or
      all of the adverse effects, including an
      explanation why the agency determined to
      pursue the action in its contemplated
      form rather than an alternative.
   4. The possible modifications to the project
      which would eliminate or minimize adverse
      effects, including a discussion of the ad-
      ditional costs involved in such modifica-
      tions.

REGULATION

1.1. Guidelines or Rules and Regulations: Interim
Guidelines prepared by the Environmental Review Board, issued June 1974; Revised Guidelines adopted November 1975.
1.2. Interpretation of Executive Order:
A. EIS Content: The requirements set forth
in the Order are further defined.
B. Threshold Criteria: Definitions of "significant" and "impact" are provided in the Guidelines. In general, any alteration or change
with regard to any part of the human or natural resources of the State that may notably
and adversely affect humans, use for humans,
for wildlife and fish populations, for sci-
entific study, or may notably and adversely
affect biotic communities is considered
grounds for EIS preparation.
C. Exemptions: None.

ADMINISTRATION

1.1. Central Coordinating Agency: An Environmental
Review Board, made up of seven agency representatives and ten members of the public, ad-
ministers and coordinates the program. A nine-
teen-member Interdepartmental Environmental Review Committee (Intercom), conducts the initial
review of EIS's for technicalities and potential
conflicts between the proposed action and agency policies.
1.2. Does the Coordinating Agency Possess the power...
A. to establish guidelines or rules and regulations? No; guidelines are presented to the
Governor for his action.
B. to review agency implementing regulations?
No, but Intercom indirectly serves this purpose since it is made up of representatives
of all nineteen State agencies.
C. to make threshold determinations? No. Originating agencies retain the responsibility
of determining whether an EIS is required.
However, a mechanism exists which allows
agencies to request a declaratory ruling
based on a brief abstract of the proposed
action.
D. to reject inadequate EIS's? Yes. The
Board and Intercom may determine that an EIS
is inadequate and cause it to be resubmitted.

2. <u>Coordination of EIS Requirement with Permit Process</u>: Not formally.
3. <u>Coordination of SEPA with NEPA</u>: An EIS prepared under NEPA fulfills the requirements of the State Executive Order unless the Board requires additional information concerning specific state issues.

PARTICIPATION

1. <u>Public Solicitation and Involvement</u>:
   A. <u>in the development of guidelines or rules and regulations</u>: A public hearing was held in 1974 before final Guidelines were adopted in November 1975.
   B. <u>in the review of environmental documents</u>: The Board maintains a list of interested citizens, citizen groups, agencies, and public media to which a monthly Environmental Impact Statement Status List and Board Agenda is mailed without charge. The Board serves as the primary public forum for citizen input.

LITIGATION

None.

NOTEWORTHY FEATURES

- Michigan's approach to impact assessment differs from NEPA and most other States in that considerations of human ecology are clearly articulated and used in evaluating environmental impacts.
- A two-step review process involving a nineteen-agency technical review Committee (Intercom) and a seventeen-member Environmental Review Board is utilized with the power of rejection being vested in the latter.
- The Michigan review process does not provide for the preparation of a draft EIS; impact statements are either accepted, accepted with recommendations, or are returned to be redone.

SEPA FACT SHEET

MINNESOTA

*The Minnesota Environmental Policy Act of 1973 Ch.
412, Laws of 1973, Minn. Stat. Ann. Ch. 116D (cum.
supp. 1974)*

> State Contact: Charles Kenow, Coordinator,
> Environmental Impact Statement Program, State
> Planning Agency, Environmental Quality Council,
> 550 Cedar St., St. Paul, MN 55101 (Phone: 612-
> 296-8254).

LEGISLATION

1. Extensiveness of Coverage: Impact statements
   are required for "any major governmental action" of the State and "any major private action of more than local significance." This
   includes actions proposed by private parties
   that may not require permits.
2. Criteria for EIS Preparation: Only "where there
   is potential for significant environmental effects" resulting from any of the above actions
   is an EIS necessary.
3. Departure from or Addition to the Language of
   NEPA: There are significant additions to NEPA
   language (see Noteworthy Features). In terms
   of EIS content, the following entries supplement the traditional five NEPA considerations:
   (a) Any direct or indirect adverse environ-
       mental, economic, and employment effects
       that cannot be avoided should the proposal
       be implemented.
   (b) The impact on State government of any Fed-
       eral controls associated with proposed ac-
       tions.
   (c) The multi-State responsibilities associated
       with proposed actions.

REGULATION

1.1. Guidelines or Rules and Regulations: "Rules
     and Regulations for Environmental Impact State-
     ments," issued by the Minnesota Environmental
     Quality Council, April 4, 1974, as amended Feb-
     ruary 13, 1977 (6 MCAR Sec. 3.021).
1.2. Interpretation of Legislation:
     A. EIS Content: The regulations expand upon
        the basic requirements set forth in the
        legislation.

276

B. Threshold Criteria: The regulations con-
    tain lists of factors to be considered for
    determining the need for an EIS according
    to the following categories:

    B. Major action. In determining whether
        an action is major, the following fac-
        tors shall be considered:
        1. Type of action;
        2. Scope of action, including size and
           cost;
        3. Location and nature of surrounding
           area;
        4. The totality of cumulative related
           actions, as defined by 6 MCAR Sec.
           3.025E.;
        5. Relation of the action to anticipat-
           ed growth and development; and
        6. Permit(s) and approval(s) required
           in addition to those of one primary,
           local agency.
    C. Local significance. In determining
        whether a major private action is of
        more than local significance, the fol-
        lowing factors shall be considered:
        1. Location of the action; and
        2. Area affected by the action.
    D. Potential for significant environmental
        effects. In determining whether an ac-
        tion has the potential for significant
        environmental effects, the following
        factors shall be considered:
        1. Type, extent, and reversibility of
           environmental effects;
        2. Cumulative potential effects of re-
           lated or anticipated future actions,
           as defined by 6 MCAR Sec. 3.025E.;
        3. The extent to which environmental
           effects are subject to mitigation by
           ongoing public regulatory authority;
           and
        4. The extent to which environmental ef-
           fects can be anticipated and con-
           trolled as a result of other environ-
           mental studies undertaken by public
           agencies or the project proposer, or
           of EIS's previously prepared on simi-
           lar actions.

C. Exemptions: The regulations define general
    exemptions for which the preparation of en-

277

vironmental documents shall not be required as well as more specific situations
where an Environmental Assessment Worksheet
(EAW) is or is not required. Agencies developing their own procedural guidelines
are encouraged to develop EAW and EIS exemption categories subject to the approval
of the Environmental Quality Council (EQC).

ADMINISTRATION

1.1. Central Coordinating Agency: An interagency
body, the Environmental Quality Council (EQC)
was assigned responsibility for administering
the EIS process. Staff for the EQC is located
in the State Planning Agency.
1.2. Does the Coordinating Agency Possess the power...
   A. to establish guidelines or rules and regulations? Yes, pursuant to MEPA.
   B. to review agency implementing regulations?
   No.
   C. to make threshold determinations? Although
   EQC formerly made all threshold determinations, the 1977 amendments to the regulations direct that the Council make the decision as to need for an EIS only if valid
   objections are made to the sponsoring agency's decision.
   D. to reject inadequate EIS's? Yes. The EQC
   reviews final EIS's for adequacy of procedure
   and policy and may cause inadequate EIS's to
   be revised and resubmitted.
   2. Coordination of EIS Requirement with Permit Process: Yes, through the "Environmental Permit
   Coordination Program" which involves a "master
   application" process for all projects involving
   more than one state permit.
   3. Coordination of SEPA with NEPA: Yes. Any part
   of a Federal EIS may be used pursuant to MEPA.
   However, any information required by the State
   Act or rules but not by NEPA must be added.

PARTICIPATION

1. Public Solicitation and Involvement:
   A. in the development of guidelines or rules
   and regulations: Numerous public hearings
   and an extensive period for public comment
   were used in the drafting of regulations.
   B. in the review of environmental documents:
   In addition to a Citizen's Advisory Committee

attached to the EQC and the required publication of an EQC Monitor containing permit
application notices; EIS preparation notices and Negative Declaration notices; notices of meetings or hearings on draft EIS's;
receipt of draft or final EIS's; and other
matters that fall within the jurisdiction of
the EQC, opportunities for public involvement are provided throughout the EIS process. These include:
- monthly meetings of the EQC are open to
  the public.
- mandatory public hearings or meetings on
  every draft EIS.
- provision for citizen petitions (with 500
  or more signatures) as a means of initiating environmental review.

LITIGATION

Negligible.

NOTEWORTHY FEATURES

- Not only is EQC empowered to require the
  revision of inadequate EIS's, it may also
  revise or modify a proposal if it determines that the action or project is inconsistent with the declarations of policy contained in the Act.
- The legislation provides for citizen petition as a means for initiating environmental assessment of a project.
- Public meetings or hearings are required
  to be held by the initiating or sponsoring
  agency as part of the draft EIS review process.
- The 1977 amendments to the Regulations establish a sliding system of charge-back procedures for recovering the costs of EIS
  preparation from private proposers.

MONTANA

Montana Environmental Policy Act of 1971, L. 1971, Rev. Code Mont., Section 69-6501, et seq. (Cum. Supp. 1973), as amended April 21, 1975 (Ch. 65, Section 69-6508 and Section 69-6509)

> State Contact: Duane Noel, Ecological Researcher, Montana Environmental Quality Council, Capitol Station, Helena, MT 59601 (Phone: 406-449-3742).

LEGISLATION

1. Extensiveness of Coverage: The EIS requirement applies to actions directly undertaken, funded, or permitted by state agencies.
2. Criteria for EIS Preparation: Impact statement preparation is required only for major actions significantly affecting the quality of the human environment.
3. Departure from or Addition to the Language of NEPA: MEPA recognizes that each person is entitled to a "healthful environment." Amendments to the law in 1975 also prescribe a system of fees to be paid by private applicants when "an application for a lease, permit, contract, license, or certificate will require an agency to compile an environmental impact statement."

REGULATION

1.1. Guidelines or Rules and Regulations: "Uniform Rules Implementing the Montana Environmental Policy Act," adopted by the Montana Commission on Environmental Quality (MCEQ), January 1976.
1.2. Interpretation of Legislation
   A. EIS Content: The following considerations are added to EIS content:
   (a) adverse and beneficial impacts of each alternative;
   (b) potential growth-inducing aspects of each alternative;
   (c) irreversible commitments of environmental resources including land, air, water, and energy resulting from each alternative;
   (d) economic and environmental benefits and

costs resulting from the proposed action and each alternative. Agencies should attempt to balance the results of their environmental assessments with their assessments of the net economic, technical, and other benefits of the proposed action and alternatives, and use all practicable means to avoid or minimize undesirable consequences for the environment;

    (e) a comparison of short-term costs and benefits from the proposed action and alternatives, with the effects on maintenance and enhancement of the long-term productivity of the environment.

B. <u>Threshold Criteria</u>: A "Preliminary Environmental Review" (PER) procedure is suggested for determining which actions significantly affect the environment. First, a categorical determination is made as to whether an action is "major" and then a checklist of environmental factors is prepared to determine significance.

C. <u>Exemptions</u>: Categorical exemptions are provided for ministerial actions, existing facilities operation and maintenance, and investigation and enforcement. No category of action may be formally designated as exempt from MEPA except through the rulemaking procedures of the Montana Administrative Procedures Act.

ADMINISTRATION

1.1. <u>Central Coordinating Agency</u>: The Environmental Quality Council (EQC) is responsible for general coordination of the program, including guideline preparation. The thirteen unpaid members of the Council are supported by a professional staff of four.

1.2. <u>Does the Coordinating Agency Possess the power</u>...

A. <u>to establish guidelines or rules and regulations</u>? EQC issues only guidelines; rules are adopted by the MCEQ.

B. <u>to review agency implementing regulations</u>? No. Agencies of the State are responsible for establishing their own procedures consistent with the general rules.

C. <u>to make threshold determinations</u>? No. Proposing agencies determine need for EIS according to PER procedures established by the rules.

D. to reject inadequate EIS's? No. Proposing agencies make the final decision as to impact statement adequacy based on review comments and/or litigation.
2. Coordination of EIS Requirement with Permit Process: EQC does not issue permits. Many agencies responsible for such activities, however, informally coordinate the two procedures if time limitations permit.
3. Coordination of SEPA with NEPA: The rules encourage the preparation of joint Federal-State impact statements.

PARTICIPATION

1. Public Solicitation and Involvement:
   A. in the development of guidelines or rules and regulations: No requirements in general guidelines; left to agency discretion.
   B. in the review of environmental documents: Impact statements are circulated, by the proposing agency, to other appropriate agencies and selected public and private groups and individuals. The need for a public hearing is also determined by the proposing agency.

LITIGATION

Negligible; although one case, Montana Wilderness Society vs. Department of Health (Beaver Creek), dealt with the issue of defects in EIS content.

NOTEWORTHY FEATURES

- Amendments to the law in 1975 explicitly extend the applicability of the EIS requirements to permitted actions by prescribing a system of applicant fees to be paid when "an application for a lease, permit, contract, license, or certificate will require an agency to compile an environmental impact statement."
- The EQC (Coordinating Agency) is in the legislative rather than the executive branch of government. This facilitates EQC's role as program "watchdog."
- Under both MEPA and the State Constitution, Montana citizens have the right to a "healthful environment."

NEW JERSEY

New Jersey Executive Order #53 (October 15, 1973)

State Contact: Lawrence Schmidt, Chief, Office of Environmental Review, Department of Environmental Protection, P.O. Box 1390, Trenton, NJ 08625 (Phone: 609-292-2662)

EXECUTIVE ACTION

1. Extensiveness of Coverage: EIS's are required of "all departments and agencies of the State" for "major construction projects."
2. Criteria for EIS Preparation: The order provides that all state funded or state sponsored construction projects in excess of $1 million or those projects less than $1 million but located in environmentally sensitive areas be subject to environmental review. Also, by virtue of specific authority granted the Department of Environmental Protection (DEP) to issue permits and other regulatory measures, it is possible for the Department to request EIS's for major projects within the private sector.
3. Departure from or Addition to the Language of NEPA: There are many significant differences between the executive order and NEPA (see Noteworthy Features). The "General Policy Guidelines" (attached to the Order), however, define EIS content in much the same way as NEPA.

REGULATION

1.1. Guidelines or Rules and Regulations: "General Policy Guidelines" are appended to the Executive Order; also, "Guidelines for the Preparation of an Environmental Impact Statement," issued by the Office of the Commissioner, Department of Environmental Protection, in 1973 and updated in February of 1974.
1.2. Interpretation of Executive Order:
   A. EIS Content: The requirements set forth in the Order are further delineated.
   B. Threshold Criteria: The determination as to whether an EIS is required is gauged by the size of the project, the type of project, or the project location (whether or not it is in an environmentally sensitive area). This

283

is set forth in the Executive Order.

C. Exemptions: None, except the Executive Order categorically exempts maintenance and repair projects.

ADMINISTRATION

1.1. Central Coordinating Agency: An Office of Environmental Review within the Department of Environmental Protection was charged with the responsibility for reviewing and commenting on all environmental documentation under Executive Order #53.

1.2. Does the Coordinating Agency Possess the power...

A. to establish guidelines or rules and regulations? The Department may issue guidelines to assist proposing agencies.

B. to review agency implementing regulations? No.

C. to make threshold determinations? Yes. The OER reviews the preliminary assessment and within thirty days makes a determination on the need for the sponsor to prepare a full EIS.

D. to reject inadequate EIS's? No. Upon receipt of the EIS, DEP makes a full review and reports its findings and recommendations to the State Planning Task Force. It is the responsibility of the State Planning Task Force to reconcile all environmental problems prior to clearing the project for construction.

2. Coordination of EIS Requirement with Permit Process: Yes. Once the environmental review of a private project has been completed, the project receives "conceptual" approval and is released to the line agencies within DEP responsible for permit issuance.

3. Coordination of SEPA with NEPA: All NEPA projects are exempt from review under the Order.

PARTICIPATION

1. Public Solicitation and Involvement:

A. in the development of guidelines or rules and regulations: None.

B. in the review of environmental documents: A "Weekly Bulletin" is published by DEP which gives notice of construction permit applications, impact statements and assessments, and public hearings.

LITIGATION

None; since the Executive Order generally refers to
actions of state agencies, disputes would be elevated
to the Governor for resolution, rather than the
Courts.

NOTEWORTHY FEATURES

- Criteria for the initiation of environmental
  assessment is defined specifically in the
  Executive Order (see above).
- Where DEP recommendations are not accepted
  by the proposing agency, it must file a
  written statement explaining its action with
  the State Planning Task Force which reconciles the differences. A project cannot proceed until such reconciliation takes place.
- The State of New Jersey has enacted separate
  Wetlands and Coastal Area Facilities Review
  (CAFRA) legislation which have independent
  EIS requirements.

NEW YORK

New York State Environmental Quality Review Act, Art. 8, New York State Environmental Conservation Law, effective June 1, 1976, as amended and added by Section 8-0117, 1976, 1977, and 1978.

State Contact: Terence Curran, Director, Office of Environmental Analysis, NYS Department of Environmental Conservation, 50 Wolf Road, Albany, NY 12233 (Phone: 518-457-2224).

LEGISLATION

1. Extensiveness of Coverage: The EIS prerequisite applies to actions directly undertaken, supported, or approved by any agency. Thus, applicability of the law is comprehensively extended to local as well as private actions requiring permits.
2. Criteria for EIS Preparation: Impact statements are required for actions which may have a "significant effect on the environment."
3. Departure from or Addition to the Language of NEPA: There are significant additions to NEPA language (see Noteworthy Features). In terms of EIS content, the following entries supplement the traditional five NEPA considerations:

   (f) mitigation measures proposed to minimize the environmental impact;
   (g) the growth-inducing aspects of the proposed action, where applicable and significant;
   (h) effects of the proposed action on the use and conservation of energy resources, where applicable and significant; and
   (i) such other information consistent with the purposes of this article as may be prescribed in guidelines issued by the commissioner pursuant to Section 8-0113 of this chapter.

REGULATION

1.1. Guidelines or Rules and Regulations: 6 N.Y.C. R.R., Part 617, effective September 1, 1976, revised January 24, 1978 and September 1, 1978 by the Department of Environmental Conservation (DEC).
1.2. Interpretation of Legislation:

A. <u>EIS Content</u>: Although no additional considerations are delineated, the scope and
purpose of the EIS is clarified:

(b) An environmental impact statement
should assemble relevant and material
facts upon which the decision is to be
made, should identify the essential issues to be decided, should evaluate all
reasonable alternatives and, on the basis
of these, should make recommendations.
In order to accomplish this, EIS's shall
be analytical and not encyclopedic.
B. <u>Threshold Criteria</u>: The regulations outline ten specific criteria for determining
what actions may have a significant effect
on the environment (see Text).
C. <u>Exemptions</u>: The regulations establish lists
of actions containing critical thresholds
according to the following classification
system:

Type I - Actions or classes of actions
that are likely to require preparation of environmental impact
statements because they will in
almost every instance have a
significant effect on the environment.

Type II - Actions or classes of actions
which have been determined not
to have a significant effect on
the environment and which do not
require environmental impact
statements under this Part.

<u>ADMINISTRATION</u>

1.1. <u>Central Coordinating Agency</u>: The Office of Environmental Analysis within the Department of
Environmental Conservation (DEC) is responsible
for implementation, coordination and discretionary review under the Act.
1.2. <u>Does the Coordinating Agency Possess the power</u>...
A. <u>to establish guidelines or rules and regula-
tions</u>? Yes, pursuant to SEQR.
B. <u>to review agency implementing regulations</u>?
Yes, for consistency with statewide rules
and regulations.
C. <u>to make threshold determinations</u>? No. Responsibility for determining need for EIS's

rests with the originating or lead agency.
D. to reject inadequate EIS's? No, although OEA does review SEQR EIS's on a discretionary basis and makes recommendations.
2. Coordination of EIS Requirement with Permit Process: Yes. An application for a permit for an action requiring an EIS is not complete until such draft statement has been filed and accepted by the lead agency as satisfactory.
3. Coordination of SEPA with NEPA: Yes. Compliance with SEQR is made in conjunction with Federal requirements in a single environmental reporting procedure. A negative declaration prepared under NEPA, however, does not constitute compliance.

PARTICIPATION

1. Public Solicitation and Involvement:
   A. in the development of guidelines or rules and regulations: Public hearings testimony and the review of submitted comments were used in the drafting of rules and regulations.
   B. in the review of environmental documents: While SEQR does not require public hearings at any stage of the review process, notice of preparation of draft and final EIS's is required to be filed in the Environmental Notice Bulletin. Notice of hearings must be filed in a newspaper of general circulation. Opportunity also exists for public involvement in the pre-EIS "scoping" process (see Noteworthy Features).

LITIGATION

None; the law has yet to be fully implemented.

NOTEWORTHY FEATURES

- SEQR has been amended to provide for phased implementation so as to allow localities the maximum leeway in establishing their programs.
- State and local agencies are also directed to compile lists of projects, within their jurisdictions, which have been approved and are therefore not subject to the environmental impact statement requirement. Lists of these grandfathered actions are to be sub-

288

mitted to the chief fiscal officer who will
certify that "substantial time, work or
money have been expended" on the projects
prior to the appropriate SEQR implementation
deadline.

- Amendments to SEQR include a "focusing"
  clause for EIS content and scope similar to
  that of California (see Text).
- The new regulations (January 1978) allow for
  a pre-application conference to determine
  areas of applicant confidentiality (see Text).
- The new regulations (January 1978) provide for
  pre-EIS "scoping" session to define EIS content and scope in a manner reminiscent of
  Massachusetts (see Text).

NORTH CAROLINA

North Carolina Environmental Policy Act of 1971 (1971 c. 1203, s.1), N.C. Gen. Stat. Ch. 113A (Cum. Supp. 1973)

> State Contact: Robert Thayer Broili, Environmental Planning Consultant, Department of Natural and Economic Resources, P.O. Box 27687, Raleigh, NC 27611 (Phone: 919-733-4984)

LEGISLATION

1. Extensiveness of Coverage: Impact statements are required only for actions involving the "expenditure of public moneys." NC-EPA does, however, allow for the inclusion of permitted actions in the EIS process by authorizing (but not mandating) localities to require EIS's for major private development projects.
2. Criteria for EIS Preparation: Only projects or programs "significantly affecting the quality of the environment" of the state require EIS preparation.
3. Departure from or Addition to the Language of NEPA: A discussion of the "mitigation measures proposed to minimize the impact" is required in addition to the traditional five NEPA entries.

REGULATION

1.1. Guidelines or Rules and Regulations: "Guidelines for the Implementation of the Environmental Policy Act of 1971," issued February 1972 and revised February 1976 by the Department of Administration.
1.2. Interpretation of Legislation:
 A. EIS Content: The guidelines largely reiterate the requirements set down in the legislation. They do, however, stress the need for a comparison of environmental as well as economic costs and benefits associated with alternative courses of action.
 B. Threshold Criteria: The following three conditions are established as indicators of "significance" (i.e., as requiring an EIS):
 (1) A proposed project or program has the the potential to degrade the quality of

the environment, to curtail the range
of the environment or to achieve short-
term to the disadvantage of long-term
goals;
(2) The possible effects of a project are
individually limited but cumulatively
considerable especially when such effects are growth-inducing; or
(3) The environmental effects of a project
will cause substantial adverse effects
on human beings either directly or indirectly.
C. Exemptions: None.

ADMINISTRATION

1.1. Central Coordinating Agency: Overall responsibility for coordination of NC-EPA was delegated
to the Department of Administration in 1972. In
reality, however, the Department of Natural and
Economic Resources (DNER) fills this role on an
informal basis.
1.2. Does the Coordinating Agency Possess the power...
A. to establish guidelines or rules and regulations? Informal guidelines only.
B. to review agency implementing regulations?
No. Agency regulations are, however, to be
in conformance with the general criteria
contained in the statewide guidelines.
C. to make threshold determinations? No. Responsibility for determining need for EIS's
rests with project originating agencies.
D. to reject inadequate EIS's? No, although
the Governor (or his designee) retains the
final decision authority on projects involving "major environmental adversities."
2. Coordination of EIS Requirement with Permit Process: Projects requiring permits which do not
involve public monies are exempt from the EIS
requirement (although localities are "authorized"
to require impact statements for major private
development projects).
3. Coordination of SEPA with NEPA: An EIS under
NEPA may be submitted in lieu of an impact
statement under NC-EPA. A negative declaration
prepared under NEPA, however, does not constitute compliance.

PARTICIPATION

1. Public Solicitation and Involvement:
A. in the development of guidelines or rules

and regulations: None.

B. in the review of environmental documents:
While public hearings are encouraged, they
are not required at any point in the review
process. Notice of preparation of draft
and final EIS's, however, must be published
in the bi-weekly North Carolina Environmental
Bulletin. Notice of hearings must be filed
in a newspaper of general circulation.

LITIGATION

Negligible; the law has largely been ignored (see
Text).

NOTEWORTHY FEATURES

- NC-EPA was originally enacted as a two-year
  experiment slated to expire in 1973. The
  Legislature voted in 1973 to extend the law
  until 1977 rather than make it permanent.
  The 1977 session again voted to extend the
  Act until August 1, 1981.
- Although the EIS requirement does not cover
  the issuance of permits where no public
  money is involved, NC-EPA does allow for the
  inclusion of private actions in the EIS process by authorizing (but not mandating) localities to require impact statements for
  major development projects.

SOUTH DAKOTA

South Dakota Environmental Policy Act, SL 1974, Ch.
245 (approved March 2, 1974), S.D. Comp. Laws 1967,
Ch. 11-1A (Supp. 1974).

State Contact: Harold Lenhart, South Dakota
Department of Environmental Protection, Foss
Building, Pierre, SD 57501 (Phone: 605-224-3351).

LEGISLATION

1. Extensiveness of Coverage: The law applies only to actions proposed or approved by state
   agencies.
2. Criteria for EIS Preparation: An EIS is required on any major action which may have a significant effect on the environment.
3. Departure from or Addition to the Language of
   NEPA: The law adds the following two entries
   to the traditional contents of impact statements under NEPA:
   (a) Mitigation measures proposed to minimize
       the environmental impact; and
   (b) The growth-inducing aspects of the proposed
       action.
   In addition, several classes of action are exempted from consideration under the Act:

   11-1A-3. Actions not subject to chapter--As
   used in this chapter, unless the context otherwise requires, "actions" do not include:
   (1) Enforcement proceedings or the exercise of
       prosecutorial discretion in determining
       whether or not to institute such proceed-
       ings;
   (2) Actions of a ministerial nature, involving
       no exercise of discretion;
   (3) Emergency actions responding to an immedi-
       ate threat to public health or safety;
   (4) Proposals for legislation; or
   (5) Actions of an environmentally protective
       regulatory nature.

REGULATION

1.1. Guidelines or Rules and Regulations: Informal
     guidelines issued by the Department of Environ-
     mental Protection (1974).

1.2. Interpretation of Legislation:
   A. EIS Content: The guidelines recommend the
      addition of the following consideration to
      EIS content:
      (a) Environmental effects of the project
          which will likely occur but that nature,
          size, duration are unknown or unknowable.
   B. Threshold Criteria: The guidelines suggest
      criteria for determining what is a "major
      action" and what has "significant environ-
      mental impacts" but defer to individual
      state agencies to prepare their own specific
      set of thresholds.
   C. Exemptions: None.

ADMINISTRATION

1.1. Central Coordinating Agency: The Department
     of Environmental Protection (DEP) serves only
     as a repository for draft and final EIS's pre-
     pared under the state law.
1.2. Does the Coordinating Agency Possess the power...
   A. to establish guidelines or rules and regula-
      tions? Yes. DEP has issued informal guide-
      lines (i.e., no statutory authority).
   B. to review agency implementing regulations?
      No.
   C. to make threshold determinations? No. Pro-
      ject originating agencies determine need for
      EIS according to their own threshold criteria.
   D. to reject inadequate EIS's? No. The Act
      does not provide authority for DEP to review
      statements for adequacy, other than meeting
      the specific requirements of the Act, or to
      enter the decision process.
   2. Coordination of EIS Requirement with Permit Pro-
      cess: Actions of an "environmentally protective
      regulatory nature" are exempted under the Act.
   3. Coordination of SEPA with NEPA: Responsibili-
      ties under the state law are waived for any ac-
      tion requiring a Federal EIS provided that such
      a statement complies with the requirements of
      SEPA.

PARTICIPATION

1. Public Solicitation and Involvement:
   A. in the development of guidelines or rules
      and regulations: None.
   B. in the review of environmental documents:
      Draft EIS's are circulated by the project

294

originating agency to other state agencies and interested members of the public. Final EIS's must be filed with DEP for at least thirty days prior to action and must include agency response to all comments.

LITIGATION

None.

NOTEWORTHY FEATURES

- Although the Act explicitly states that an EIS is required for actions proposed or approved by state agencies, it also exempts several classes of action including "actions of an environmentally protective regulatory nature." Thus, permitted actions are exempted despite the fact that they constitute agency "approval."

TEXAS

*Administrative action: Policy for the Environment,
adopted by the Interagency Council on Natural Resources and the Environment on March 7, 1972, and
published in "Environment for Tomorrow: The Texas
Response," subsequently updated by "The Environment:
Policy--Guidelines and Procedures for Processing
EIS's," developed and adopted by the Interagency
Council on Natural Resources and the Environment,
published November 1975.*

State Contact: Charles D. Travis, Director,
Governor's Budget and Planning Office, Exec-
utive Office Building, 411 W. 13th Street,
Austin, TX 78701 (Phone: 512-475-6156).

ADMINISTRATIVE ACTION

1. Extensiveness of Coverage: The EIS suggestion
   applies to all projects proposed, funded, or
   approved by member agencies of the Governor's
   Interagency Council on Natural Resources and
   the Environment.
2. Criteria for EIS Preparation: An EIS is suggested only for "project proposals significantly
   affecting the quality of the human environment."
3. Departure from or Addition to the Language of
   NEPA: It is suggested that EIS's contain, in
   addition to NEPA's five requirements, consideration of:
   (a) the expected beneficial environmental im-
       pacts of the proposed actions; and
   (b) the appropriate extent to which the pro-
       posed action will affect the supply and
       conservation of energy in the state as well
       as a statement of energy efficiency.

REGULATION

1.1. Guidelines or Rules and Regulations: "The En-
     vironment; Policy--Guidelines and Procedures
     for Processing EIS's" developed and adopted by
     the Interagency Council on Natural Resources
     and the Environment (ICNRE), in November 1975.
1.2. Interpretation of Administrative Action:
     A. EIS Content: The format of environmental
        factors to be considered is clarified.
     B. Threshold Criteria: None, although a

detailed initial assessment process involving the comparison of social, economic, and
environmental factors with and without the
project is established.
C. Exemptions: Agencies are asked to categorize projects or activities as follows:
Type I   Projects - will always require
                     EIS's
Type II   Projects - may or may not require
                     EIS's depending on the
                     individual significance
                     of the projects.
Type III Projects - will not require EIS's.

ADMINISTRATION

1.1. Central Coordinating Agency: The Interagency
Council on Natural Resources and the Environment adopted a voluntary, cooperative policy
for preparing EIS's among its member state
agencies. The Division of Planning Coordination (DPC) coordinates the review of EIS's
for the originating member agency.
1.2. Does the Coordinating Agency Possess the power...
A. to establish guidelines or rules and regulations? Yes; self-imposed by ICNRE.
B. to review agency implementating regulations?
No.
C. to make threshold determinations? No.
D. to reject inadequate EIS's? No.
2. Coordination of EIS Requirement with Permit Process: None.
3. Coordination of SEPA with NEPA: Any projects
involving Federal funds are exempt from the
State Administration Act.

PARTICIPATION

1. Public Solicitation and Involvement:
A. in the development of guidelines or rules
and regulations: None. Guidelines were
adopted as a matter of policy.
B. in the review of environmental documents:
None.

LITIGATION

None.

NOTEWORTHY FEATURES

- The EIS process adopted by the Interagency Council on Natural Resources and the Environment is voluntary and contains no mandates or requirements of its member agencies.
- The guidelines stress the comparison of "without-project" conditions and "with-project" effects in preliminary assessment.
- The guidelines suggest a three-level system of project categorization:

  Type I   - will always require EIS's
  Type II   - may or may not require EIS's depending on the individual significance of the projects
  Type III   - will not require EIS's

UTAH

State of Utah Executive Order, August 27, 1975.

> State Contact: Chauncey Powis, Environmental
> Coordinator, State Planning Coordinator's Of-
> fice, 118 State Capitol, Salt Lake City, UT
> 84114 (Phone: 801-533-5246).

EXECUTIVE ACTION

1. Extensiveness of Coverage: The EIS requirement
   applies to state agencies proposing or having
   administrative responsibility for actions (i.e.,
   permitted actions).
2. Criteria for EIS Preparation: An EIS is required only when an action "has the potential
   significantly to affect the environment."
3. Departure from or Addition to the Language of
   NEPA: The Executive Order adds the consideration of "mitigative measures included in the
   proposed action" to the required contents of
   EIS's.

REGULATION

1.1. Guidelines or Rules and Regulations: "Guide-
     lines for Implementation of the Executive Order
     on Environmental Quality," promulgated by the
     Environmental Coordinating Committee in 1974.
1.2. Interpretation of Executive Order:
     A. EIS Content: The requirements set forth in
        the Order are further defined.
     B. Threshold Criteria: The guidelines clarify
        the meaning of "significant environmental
        impact." Consideration of cumulative ef-
        fects, public controversy, irreversibility of
        impact, duration of impact,and scope and
        stability of affected ecosystems are stressed
        in determining significance.
     C. Exemptions: Although the guidelines specify
        no exemptions, agencies are encouraged to
        submit individual lists of actions they would
        like to have exempted to the Environmental
        Coordinating Committee for review.

ADMINISTRATION

1.1. Central Coordinating Agency: An Environmental
     Coordinating Committee (ECC), consisting of

nineteen agency representatives, is established
by the Order as the central review and coordinating body; ECC is an advisory committee to
the Economic and Physical Development Interdepartmental Coordination Group (EPDICG) on environmental issues.

1.2. Does the Coordinating Agency Possess the power...
   A. to establish guidelines or rules and regula-
      tions? Yes, pursuant to the executive order.
   B. to review agency implementing regulations?
      No, but ECC serves to review agency requests
      for exempt actions.
   C. to make threshold determinations? Yes, through
      an environmental assessment review process in-
      volving ECC and EPDICG, the Governor retains
      the ultimate power to require the preparation
      of a full EIS.
   D. to reject inadequate EIS's? Yes, through
      the process outlined in (C).

2. Coordination of EIS Requirement with Permit Process: Yes, informally.
3. Coordination of SEPA with NEPA: Projects requiring EIS's under NEPA are exempt from the
   Executive Order.

PARTICIPATION

1. Public Solicitation and Involvement:
   A. in the development of guidelines or rules
      and regulations: None.
   B. in the review of environmental documents:
      Although the Order calls for the distribu-
      tion of documents to public, no specific
      mechanisms are established to ensure public
      notification or participation.

LITIGATION

None.

NOTEWORTHY FEATURES

   • Utah uses a three-step review process; en-
     vironmental documents are first submitted
     to the Office of the State Planning Coordi-
     nator (State Clearinghouse) for transmittal
     to and review by the Environmental Coordina-
     ting Committee. Unless the action is unani-
     mously approved by ECC, it is submitted to
     the Economic and Physical Development Inter-
     departmental Coordination Group for consid-

eration. If the project is disapproved or
approved by a divided vote, it must be submitted to the Governor for final action.

- Projects are allowed to proceed without review agency approval if comments are not received within definite and stringent time
limits (thirty days for state agencies;
forty-five days for local and Federal
agencies).

VIRGINIA

Virginia Environmental Quality Act of 1972, Ch. 774, Laws of 1972, Va. Code Ann. Ch. 17, Sections 10-177-10-186 as amended by Ch. 354, Laws of 1974; Va. Environmental Impact Reports Act of 1973, Ch. 384, Laws of 1973, Va. Code Ann. Ch. 1.8, Sections 10-17.107 - 10-17.112, as amended by Ch. 404. Acts of Assembly, 1977.

State Contact: Reginald Wallace, Environmental Impact Statement Coordinator, Virginia Council on the Environment, 903 Ninth Street Office Building, Richmond, VA 23219 (Phone: 804-786-4500).

LEGISLATION

1. Extensiveness of Coverage: The EIS requirement applies to "all State agencies, boards, authorities, and commissions or any branch of the State government."
2. Criteria for EIS Preparation: Impact statements are required only for "major" state projects. This is further defined as the acquisition of land for any state facility construction, the construction of any facility or expansion of an existing facility which is hereafter undertaken by any state agency, board, commission, authority or any branch of the State government, including state-supported institutions of higher learning, which costs one hundred thousand dollars or more; provided, this term shall not apply to any highway or road construction or any part thereof.
3. Departure from or Addition to the Language of NEPA: EIS's are to include the following:
   (1) The environmental impact of the proposed construction;
   (2) Any adverse environmental effects which cannot be avoided if the proposed construction is undertaken;
   (3) Measures proposed to minimize the impact of the proposed construction;
   (4) Any alternatives to the proposed construction; and
   (5) Any irreversible environmental changes which would be involved in the proposed construction.

REGULATION

1.1. Guidelines or Rules and Regulations: "Procedures Manual and Guidelines for the Environmental Impact Statement Program in the Commonwealth of Virginia," issued by The Council on
the Environment in June of 1976.
1.2. Interpretation of Legislation:
   A. EIS Content: The guidelines define the re-
      quired information largely according to the
      format set down in the legislation. Addi-
      tional criteria for PEIS's, however, is pro-
      vided by certain individual agencies when-
      ever a project will impact a specific area
      of concern (e.g., water, air, marine re-
      sources, and forestry).
   B. Threshold Criteria: Criteria for determin-
      ing what actions require the preparation of
      EIS's is set by law (above).
   C. Exemptions: None. No class of action is
      categorically exempted other than that set
      forth in the legislation.

ADMINISTRATION

1.1. Central Coordinating Agency: The Council on the
Environment (COE) is designated by the legislation as responsible for coordinating the circulation and review of impact statements as well
as the development of criteria and procedures
for the EIS process.
1.2. Does the Coordinating Agency Possess the power...
   A. to establish guidelines or rules and regula-
      tions? Yes, pursuant to the impact statement
      legislation.
   B. to review agency implementing regulations?
      No.
   C. to make threshold determinations? No. Each
      agency in the review process decides if a
      project will have significant impacts al-
      though COE does screen "Forms of Intent" to
      estimate caseload.
   D. to reject inadequate EIS's? No, although
      approval of the Governor is required before
      funds for a proposed action may be appropri-
      ated.
   2. Coordination of EIS Requirement with Permit Pro-
      cess: Although the EIS process does not apply
      to permitted actions, amendments to House Bill
      962 in 1976 strengthened the Council's role with
      respect to multiple permit coordination. Rules,

regulations, and procedures have been adopted
and are in effect.
3. Coordination of SEPA with NEPA: EIS's prepared
   under NEPA are acceptable so long as they contain the considerations specified in the State
   Guidelines.

PARTICIPATION

1. Public Solicitation and Involvement:
   A. in the development of guidelines or rules
      and regulations: None.
   B. in the review of environmental documents:
      The Council issues a widely distributed
      monthly publication listing all EIS's, both
      state and Federal, that are currently avail-
      able for review; the Council maintains a
      complete lending library of environmental
      documents. Agencies must be responsive to
      review comments.

LITIGATION

None.

NOTEWORTHY FEATURES

- Virginia separates its statement of environmental policy (Ch. 774) from its environmental impact statement requirement (Ch. 384).
  Thus, the role of the Council on the Environment (established under VEQA) was sol-
  idified before the responsibility for coordinating the state EIS law was added.
- The criteria for threshold determination and
  categorical exemption are set forth in the
  legislation of Virginia's EIS requirement:
  (b) "Major State project" means the acquisition of land for any State facility construction, the construction of any facility or
  expansion of any existing facility which is
  hereafter undertaken by any state agency,
  board, commission, authority or any branch
  of the state government, including state-
  supported institutions of higher learning,
  which costs one hundred thousand dollars or
  more; provided, this term shall not apply to
  any highway or road construction or any part
  thereof. For the purposes of this chapter,
  authority shall not include any industrial
  development authority created pursuant to

the provisions of chapter 33 of Title 15.1
of this Code or chapter 643, as amended, of
the 1964 Acts of Assembly. Nor shall authority include any housing development or redevelopment authority established pursuant to
state law. For the purposes of this chapter,
branch of the state government shall not be
construed to include any county, city or town
of the Commonwealth.

- In January of every odd-numbered year a Form
of Intent is required by the Council from
each state agency proposing capital outlay
projects for which funds will be requested
for the subsequent biennium. Through this
device the Council can anticipate the number
of PEIS's to be submitted, for what projects,
and when to expect them. This information
then becomes available to the review agencies
as well as to all other interested persons.
The primary benefit is that meaningful interactions can begin between the agency
sponsoring the projects and all interested
parties early in the planning process and,
therefore, modifications can be made to the
proposed facilities, if necessary, prior to
the preparation of the PEIS.

- While NEPA requires only that agencies include EIS's in recommendations or reports,
the Environmental Impact Reports Act establishes a procedure for incorporating the
EIS into the decision-making process.

WASHINGTON

State Environmental Policy Act of 1971, Rev. Code Wash. Ch. 43.21C (Supp. 1973), as amended by Sub. Senate Bill 3277, Ch. 179, Laws of 1974.

> State Contact: Stan Springer, Environmental Review Section, State of Washington Department of Ecology, Olympia, WA 98504 (Phone: 206-753-1654).

LEGISLATION

1. Extensiveness of Coverage: All branches of government are subject to the Act, including "State agencies, municipal and public corporations, and counties." The EIS requirement has also been comprehensively extended to actions requiring permits at any level of government.
2. Criteria for EIS Preparation: Only "major actions significantly affecting the quality of the environment" require impact statements.
3. Departure from or Addition to the Language of NEPA: Although the original legislation was a virtual duplicate of NEPA, the 1974 amendments provided significant additions (see Noteworthy Features).

REGULATION

1.1. Guidelines or Rules and Regulations: "Guidelines for Implementation of the State Environmental Policy Act of 1971," prepared by the Department of Ecology, as revised by "State Environmental Policy Act Guidelines" (WAC 197-10), issued by the Council on Environmental Policy, January 16, 1976, as revised December 1977 by the Department of Ecology.
1.2. Interpretation of Legislation:
   A. EIS Content: Aside from expanding upon the requirements in the legislation, the guidelines also add a discussion of "adverse environmental impacts which may be mitigated" as well as consideration of "direct and indirect effects" and "the relationship between the costs of the unavoidable adverse impacts and the expected beneficial environmental impacts."
   B. Threshold Criteria: An "Environmental Check-

list Form" has been developed for determining need for an EIS. Criteria include a spectrum of natural as well as socioeconomic considerations.

C. Exemptions: A significant portion of the Guidelines is devoted to categorical exemptions. Unlike most other such exemptions, however, Washington's set down very specific criteria that avoid arbitrary designations such as "ministerial actions."

ADMINISTRATION

1.1. Designation of Central Coordinating Agency: The Department of Ecology is responsible for coordinating the SEPA program, including the issuance of implementing Guidelines.

1.2. Does the Coordinating Agency Possess the power...

A. to establish guidelines or rules and regulations? Yes. Pursuant to SEPA, state and local agencies must adopt their own regulations consistent with the Statewide Guidelines.

B. to review agency implementing regulations? Yes.

C. to make threshold determinations? No. Project originating or lead agencies determine need for EIS according to an established "Environmental Checklist Form" issued by DOE.

D. to reject inadequate EIS's? No, although the condition and denial of permits constitutes a form of enforcement authority for private activities.

2. Coordination of EIS Requirement with Permit Process: Yes, through an Environmental Coordination Procedures Act.

3. Coordination of SEPA with NEPA: An adequate EIS prepared under NEPA may be used in lieu of one under SEPA. Negative declarations prepared under NEPA, however, are not acceptable.

PARTICIPATION

1.1. Public Solicitation and Involvement:

A. in the development of guidelines or rules and regulations: Several thousand copies of the successive drafts were broadly distributed. Public hearings were held throughout the State. Lengthy public review and comment periods were included.

B. in the review of environmental documents:

Lead agencies are responsible for sending
all declarations of non-significance, draft
and final EIS's to DOE. The Department is
then responsible for the weekly listing and
distributing of these documents in a "SEPA
Register." Furthermore, a public hearing
may be initiated when fifty or more persons
adversely affected by a proposal make a
written request to the lead agency within
thirty-five days of the issuance of the draft
EIS.

LITIGATION

Extensive; second only to California. Between 1972
and 1976, this included ten Supreme Court cases involving the same range of issues as NEPA.

NOTEWORTHY FEATURES

- Amendments to the law in 1974 mandated both
  the creation of a temporary rule-making body
  for SEPA (The Council on Environmental Policy) and the initiation of a comprehensive
  study of SEPA implementation status.
- Amendments to SEPA in 1976 established a
  statute of limitation concerning environmental review and legal action.
- The system of categorical exemptions developed in the Guidelines is very specific
  and does not include provisions for blanket
  exemptions of designations such as "ministerial actions."
- DOE has established an "Environmental Impact
  Information System" which categorizes and
  summarizes all EIS information received by
  the agency for future use.
- When fifty or more persons affected by the
  impact of a proposed project or two or more
  agencies with jurisdiction make written request to the lead agency, a public hearing is
  required.
- The guidelines encourage each locality to
  designate environmentally sensitive areas
  within their jurisdiction within which the
  general categorical exemptions would not
  apply.

WISCONSIN

Wisconsin Environmental Policy Act of 1971, Ch. 174, Laws of 1971, adding Wisc. Stat. Ann. Ch. 1, Section 1.11 et seq. (Cum. Supp. 1074-75) as amended 1973; also Chapter 273, Section 23.11(5) Laws of 1971.

LEGISLATION

1. **Extensiveness of Coverage:** The law requires all "agencies of the state" to prepare EIS's on proposals for legislation and for other "major actions" (interpreted to include management and regulatory functions of agencies).
2. **Criteria for EIS Preparation:** Under Chapter 274, an EIS is required only for major actions "significantly affecting the quality of the human environment."
3. **Departure from or Addition to the Language of NEPA:** WEPA contains two additional requirements:
   (a) each impact statement shall also contain details of the beneficial aspects of the proposed project, both short term and long term, and an analysis of the economic advantages and disadvantages of the proposal; and
   (b) a public hearing shall be held on each proposal for which an EIS was prepared.

REGULATION

1.1. **Guidelines or Rules and Regulations:** "Revised Guidelines," issued by Governor's Executive order No. 26 (February 1976).
1.2. **Interpretation of Legislation:**
   A. **EIS Content:** The guidelines do not add any new areas of consideration but greatly define those set down in the legislation.
   B. **Threshold Criteria:** The guidelines define what constitutes an "action" under WEPA as well as establishing criteria for determining whether an action is "significant." A screening worksheet procedure is installed as the mechanism for determining the need for an EIS.
   C. **Exemptions:** The guidelines require that agencies categorize their actions according to the following three types:

      (1) Type I actions which will always require
      an EIS.
      (2) Type II actions which may or may not re-
      quire impact statements depending on the
      individual significance of a project; and
      (3) Type III actions which will never require
      EIS's.
     These lists require periodic revision subject
     to public scrutiny and comment.

     Regulatory actions taken by the Department of
     Natural Resources are specifically excluded.

ADMINISTRATION

1.1. Central Coordinating Agency: An Inter-Agency
Environmental Impact Statement Coordinating Committee (composed of representatives of all thirty major state agencies) is responsible for coordinating the implementation of the Act and
the maintenance of guidelines.
1.2. Does the Coordinating Agency Possess the power...
   A. to establish guidelines or rules and regulations? Yes, through Executive Order of the
   Governor.
   B. to review agency implementing regulations?
   No, but the guidelines set specific requirements for state agencies.
   C. to make threshold determination? No. Project originating agencies determine need for
   EIS according to procedures established in
   the guidelines.
   D. to reject inadequate EIS's? No, although
   the Interagency Coordinating Committee
   serves as a "sounding board" for questions
   relating to compliance.
  2. Coordination of EIS Requirement with Permit Process: Yes.
  3. Coordination of SEPA with NEPA: In cases where
  a Federal EIS is required, no separate state
  EIS is necessary providing the agency finds the
  Federal EIS satisfactory.

PARTICIPATION

1. Public Solicitation and Involvement:
   A. in the development of guidelines or rules
   and regulations: Public comment.
   B. in the review of environmental documents:
   The guidelines require agencies to prepare
   a preliminary environmental report (PER)

similar to a draft impact statement; PER's
must be made available to the public for a
forty-five day review period. Comments received are then used in the development of
an EIS. The EIS is circulated for a thirty
day review followed by a mandatory public
hearing. Proposed changes in administrative
process would require participation in the
screening process leading to an EIS decision.

LITIGATION

Moderate; there have been two recent (1977) Supreme
Court cases, one of which (Wisconsin Environmental
Decade v. Pub. Service Comm., 79 Wis. 2d 161) decided
that Federal law interpreting similar NEPA provisions
is "persuasive authority" in interpreting WEPA. The
other (Wisconsin Environmental Decade v. Pub. Service Comm. and Wisc. Electric Power Co., 79 Wis. 2d
409) stressed that the burden of justifying a negative EIS determination is on the agency. The question of the role of EIS's in regulatory decisions is
replacing EIS threshold decisions as the key issue.

NOTEWORTHY FEATURES

- While WEPA does not specifically require an
  EIS for private activities, a companion law
  (Chapter 273, Sec. 23.11(5)) provides a
  mechanism for the Department of Natural Resources to request EIS's from applicants
  seeking Departmental permits or statutory
  approvals, provided that the area affected
  is more than forty acres in extent or the
  estimated cost of the project exceeds
  $25,000.
- Every EIS must contain details concerning
  the economic advantages and disadvantages
  of the proposal.
- Agencies are required to categorize their
  actions according to three types:
  (1) Type I actions which will always require
      EIS's;
  (2) Type II actions which may or may not re-
      quire EIS's depending on the individual
      significance of a project; and
  (3) Type III actions which will never re-
      quire EIS's.
- A public hearing is mandatory for every proposal requiring an EIS (other than legislation) before a final decision is made.

- The Department of Natural Resources maintains an EIS library and has developed a data process system for storage and retrieval of environmental data.

Appendix B
Addenda to Chapters 4 and 5

COMPLETED CEQA DOCUMENT COSTS IN RELATION TO PROJECT COSTS
(includes departments which have reported as of 12/9/75)
FY 73 thru FY 75 unless otherwise noted

ORGANIZATION	PROJECT COSTS	DOCUMENT COSTS	DOCUMENT COSTS AS PERCENT OF PROJECT COSTS
BUSINESS AND TRANSPORTATION			
*California Highway Patrol	$8,034,084 (10 projects, all field offices	$16-$52 @ (all very simple negative declarations	.001%
*Dept. of Motor Vehicles	$1,125,000 (precise costs available for only 1 project)	$2,000(E.I.R.)	.18%
*CalTrans FY 71 thru FY 75	Estimate of avg. cost per mile to const. in <u>1975</u>	Avg. Doc. cost per mile of facility (sample)	
	Metro Frwy 12,000,000	Metro Frwy $150,000	.04%-1.7%
	Urban Frwy 6,000,000	Urban Frwy $60,000	.04%-1.7%
	Suburban Frwy 6,000,000	Suburban Frwy $30,000	.01%-.06%
	Rural Hwy 1,500,000	Rural Hwy $12,000	.01%-2.5%
*Dept. of Real Estate	While D.R.E. acts as a lead agency, it does not actually prepare the environmental documents. Documents are prepared by proponents and the Department keeps no record of their expenditures.		
HEALTH AND WELFARE AGENCY			
*Dept. of Health	$120,000 (one project)	$7,500 (one EIR)	6.25%
*Employment Development Dept.	$3,703,450 (8 projects)	$ 900 (all Neg. Dec.)	.02%
RESOURCES			
*Parks & Rec.	not available	$186,000	not available
*Navigation & Ocean Develop.	$1,561,000 (15 projects)	$ 8,900 (12 EIRs & 3 Neg.Dec.)	.57%
*Water Resource Control Board	Projects are applications for diversion of water. No capital outlay is involved.		
*Dept. of Water Resources			.04%-2.56% (for 10 projects)
*Dept. of Gen. Services	$1,700,000 (one project)	$50,000 (one EIR)	2.9%

TABLE B
COST OF REVIEW OF ENVIRONMENTAL DOCUMENTS
Estimated Expenditures
(in thousands of dollars)
FOR FISCAL YEAR 1975-76

ORGANIZATION	ALL REVIEW ACTIVITY	NEPA DOCUMENT REVIEWS	CEQA DOCUMENT REVIEWS
Agriculture & Services Agency			
*Dept. of Agriculture	26	5	21
Business & Transportation Agency			
Dept. of Transportation	210	50	160
Health & Welfare Agency			
*Dept. of Health	18	4	14
Resources Agency			
*Air Resources Board	157	25	95
*Bay Conservation and Development	62	31	31
*Coastal Zone Conservation Commission	102	76	26
*Dept. of Conservation	171	10	112
*Dept. of Fish & Game	1,239	130	519
*State Lands Division	122	73	49
*Dept. of Navigation & Ocean Development	15	1	7
*California Pollution Financing Authority	39	4	35
*Dept. of Parks and Recreation	90	32	58
*Reclamation Board	30	15	15
*Solid Waste Management	72	7	65
*Dept. of Water Resources	329	71	212
*Water Resources Control Board	417	21	191
TOTAL	2,845	425	1,415

UNIT	GUIDELINE DEVELOPMENT				IMPACT ASSESSMENT	
	No. of Projects	Staff Assigned	Staff Hours	Estimated Costs	No. of Projects	Staff Assigned
STATE AGENCIES						
Eastern Washington State College	4	2	20	500	5	6
Evergreen State Coll.		2	135	1,516	6	2
Washington State Univ.			72	900	19	
Central Washington State College		2	80	880	20	4
Univ. of Washington	2	36	550	4,920	87	22
Western Washington State College		2	40	400	24	6
TOTAL	635	135	14,889	173,133	9,548	333
State Library						
Military Dept.	1	2	80	420	2	2
Dept. of Motor Vehicles		5	122	1,507	332	6
Dept. of Nat. Resources	2	3	560	17,864	2,528	68
Oceanographic Commission						
OPP&FM		2	8	72		4
Parks & Rec. Commission	1	4	800	6,000	891	18
Plng. & Comm. Affairs Agency		3	313	3,556		11
Pollution Control Hearing Bd						
Dept. of Social & Health Services	1	23	446	5,920	432	56
State Capitol Committee						
State Capitol Hist. Society						
Super. of Public Instruction						
Toll Facilities						
Traffic Safety Comm.						
Urban Arterial Bd.		1	30	600		
Utilities & Transportation Comm.		5	1,657	32,241		8
Washington State Hist. Association						

STATE ENVIRONMENTAL POLICY ACT
ENVIRONMENTAL IMPACT PROCEDURES
(continued)
STATEMENT
PREPARATION

Staff Hours	Estimated Costs	No. of Projects	Staff Assigned	Staff Hours	Estimated Costs	Total Est. Cost
140	3,700	4	4	400	1,500	5,700
28	319					1,835
422	5,150	19		206	27,350	33,400
80	880					1,760
1,392	30,920	6	24	6,282	53,780	89,620
270	2,800		3	2,000	18,500	21,700
82,831	1,552,186	760	352.6	332,152	4,246,402	5,971,721
32	192					612
455	2,647					4,154
6,535	303,636	4	8	550	43,974	365,474
1,252	9,808					9,808
2,770	61,400	20	3	6,198	46,800	114,200
1,758	13,048					16,604
1,305	12,750	8	4	185	6,400	25,070
						600
214	4,491					38,732

(continued)

UNIT STATE AGENCIES	GUIDELINE DEVELOPMENT				IMPACT ASSESSMENT	
	No. of Projects	Staff Assigned	Staff Hours	Estimated Costs	No. of Projects	Staff Assigned
Eastern Washington State College	4	2	20	500	5	6
Evergreen State College		2	135	1,516	6	2
Washington State University			72	900	19	
Central Washington State College		2	80	880	20	4
University of Washington	2	36	550	4,920	87	22
Western Washington State College		2	40	400	24	6
TOTAL	635	135	14,889	173,133	9,548	333

STATEMENT
PREPARATION

Staff Hours	Estimated Costs	No. of Projects	Staff Assigned	Staff Hours	Estimated Costs	Total Est. Cost
140	3,700		4	400	1,500	5,700
28	319					1,835
422	5,510	19		206	27,350	33,400
80	880					1,760
1,392	30,920	6	24	6,282	53,780	89,620
270	2,800		3	2,000	18,500	21,700
82,831	1,552,186	760	352.6	332,152	4,246,402	5,971,721

UNIT RESPONDING		Department of Ecology	No. of Projects	No. of Staff Assigned
IMPLEMENTATION	GUIDELINE PREPARATION	Dept. SEPA rules-- WAC 173-800	1	.5
	OTHER (SPECIFY)	Occasions advising/assisting local agencies on SEPA(est)	600	.9 (continuing)
		Development of a model ordinance, per legislation, for optional local use	1	.5
		*Acting on consultation requests and transfers of lead agency responsibility	15	.4 (continuing)
ADMINISTRATION	EIS PREPARATION	*EIS on actions proposed by the department. Includes new categories where DOE is automatically lead agency.	37	1.6 (continuing)
	EIS REVIEW	Includes an estimated 90 NEPA statements	670	.6 (continuing)
	OTHER (SPECIFY)	*Guideline questions to the dept. since assignment of Council duties, 7/1/76	420	.8 (continuing)
		*Responding to petitions for administrative changes in guidelines since 7/1/76.	15	.6 (continuing)
OTHER ACTIVITY	SPECIFY- (i.e., INFORMATION DISSEMINATION MEETINGS, PLANNING PROCESS, ETC.)	Applicant's brochure on SEPA	1	.4
		Dept. meetings with all state agencies	3	.1
	*Maint. register and respond to requests on SEPA statutory notices		550	.2 (continuing)

ACTIONS TAKEN TO IMPLEMENT SEPA:
Implementation of SEPA has progressed but has been limited in
the department due to budget and staff constraints in those
duties delegated solely to the department.

COMMENTS - (i.e., VALUE OF PROCESS SUGGESTIONS FOR IMPROVEMENT, ETC.)
The SEPA process is showing value and effectiveness in the following
ways:
1 - Identification of lead agency and a decision maker on proposals.
2 - Public availability of environmental information.
3 - Problem solving and impact reduction in projects - frequently
    prior to any need for an EIS.
4 - Encouragement of interagency coordination on proposals.
The department is additionally studying potential revisions to
simplify and clarify the guidelines.

Staff Hours (Time Involved)	Total Estimated Cost
200	$2,000
32/wk.	320/wk
300	3,000
16/wk.	160/wk.
64/wk.	640/wk.
24/wk.	240/wk.
32/wk.	320/wk.
24/wk.	240/wk.
200	2,000
30	300
8/wk.	80/wk.

NOTE: *Indicates those new workload items assigned uniquely
to the department by the SEPA guidelines (Council on
Environmental Policy). Supplemental staff and budget
are being requested to accomplish these new duties
and to reduce the impact to other priority activities
in the department.

A listing of Type I and Type II actions and classes of actions follows:

TYPE I ACTIONS

Actions or classes of actions which are likely to, but will not necessarily, require preparation of environmental impact statements because they will in almost every instance have a significant effect on the environment:

(a) Construction of new (or expansion by more than 50% of existing size, square footage or usage of existing):
(1) Airports
(2) Facilities such as hospitals, schools, institutions of higher learning, correction facilities, and major office centers
(3) Road or highway sections (including bridges) which require an indirect source permit under 6 NYCRR Part 203
(4) Dams with a downstream hazard of "C" classification under Environmental Conservation Law (ECL) section 15-1503
(5) Chemical pulp mills
(6) Portland cement plants
(7) Iron and steel plants
(8) Primary aluminum ore reduction plants
(9) Sulfuric acid plants
(10) Petroleum refineries
(11) Lime plants
(12) Bi-product coke manufacturing plants
(13) Sulfur recovery plants
(14) Fuel conversion plants
(b) Construction of new; expansion either by 50% or by an increment equal to or in excess of any of the numerical thresholds listed below:

(1) Parking facilities or other facilities with an associated parking area of 250 or more cars only if such facility would require an indirect source permit under 6 NYCRR Part 203
(2) Stationary combustion installations operating at a total heating input exceeding 1,000 million BTU's per hour
(3) Incinerators operating at a refuse charging rate exceeding 250 tons of refuse per 24 hour day
(4) Storage facilities designed for or capable of storing one million or more gallons of liquefied natural gas, liquefied petroleum gas or other liquid fuels
(5) Process, exhaust and/or ventilation systems

emitting air contaminants assigned an environmental rating of "A" under 6 NYCRR 212 and whose total emission rate of such "A" contaminants exceeds 1 pound per hour

(6) Process, exhaust and/or ventilation systems from which the total emission rate of all air contaminants exceeds 50 tons per day

(7) A sanitary landfill for an excess of 100,000 cubic yards per year of waste fill

(8) Any facility, development or project which would generate more than 5,000 vehicle trips per any hour or more than 25,000 vehicle trips per any eight-hour period

(9) Any facility, development or project which would use ground or surface water in excess of 2,000,000 gallons in any day

(10) Any industrial facility, which has a yearly average discharge flow, based on days of discharge, of greater than 500,000 gallons per day

(11) Any publicly or privately owned sewage treament works which has an average daily design flow of more than 500,000 gallons per day

(12) A residential development outside any standard metropolitan statistical area as defined by the U.S. Census Bureau that includes 50 or more units in an unsewered area or 250 or more units in a sewered area or within a standard metropolitan statistical area that includes 50 or more units in an unsewered area or 2,500 or more units in a sewered area (see definitions of "unsewered area" and "residential development" below)*

(13) Lakes or other bodies of water with a water surface in excess of 200 acres

(c) Any facility, development or project which is to be directly located in one of the following critical areas:

(1) tidal wetlands as defined in Article 25 of the ECL

(2) freshwater wetlands as defined in Article 24 of the ECL

(3) areas of special flood hazard (flood plain areas having special flood hazards) as defined in 6 NYCRR 500

(4) wild, scenic and recreational river areas designated in Title 27 of Article 15 of the ECL and areas designated as wild and scenic rivers under provisions of P.L. 90-542, the National Wild and Scenic Rivers Act.

---

*An unsewered area means an area proposed for residential development where sewage disposal will not be to an existing municipal sewerage system and treatment works. Residential development shall include any realty subdivisions, mobile home parks, travel vehicle parks and campsites.

(d) Any facility, development or project having an adverse impact on any historic or prehistoric site,
building, or structure listed on the National Register
of Historic Places or having an adverse impact on any
historic or prehistoric building, structure or site
that has been formally proposed by the Committee on
the Registers for consideration by the New York State
Board on Historic Preservation for a recommendation to
the State Historic Officer for nomination for inclusion
in said National Register

(e) Any development, project or permanent facility of
a non-agricultural use in an agricultural district
except those listed as Type II actions

(f) Application of pesticides or herbicides over more than
1500 contiguous acres

(g) Clearcutting of 640 or more contiguous acres of forest
cover or vegetation other than crops

(h) Adoption of comprehensive land use plans, zoning
ordinances, building codes, comprehensive solid waste
plans, state and regional transportation plans, water
resource basin plans, comprehensive water quality
studies, area-wide waste water treatment plans, state
environmental plans, local flood plain control plans,
and the like

(i) Commercial burial of radioactive materials requiring
a permit under 6 NYCRR Part 380

(j) Any action which will result in excessive or
unusual noise or vibration, taking into consideration
the volume, intensity, pitch, time duration and the
appropriate land uses for both the source and the
recipient of such noise

(k) Acquisition or sale by a public agency of more
than 250 contiguous acres of land

TYPE II

Actions or classes of actions which have been determined
not to have a significant effect on the environment and do not
require environmental impact statements under this Part:

(a) Construction or alteration of a single or two-family
residence and accessory appurtenant uses or structures not in conjunction with the construction or
alteration of two or more such residences and not
in one of the critical areas described in this section
for Type I actions

(b) The extension of utility facilities to serve new
or altered single or two family residential structures
or to render service in approved subdivisions

(c) Construction or alteration of a store, office or restaurant designed for an occupant load of 20 persons or less, if not in conjunction with the construction or alteration of two or more stores, offices or restaurants and if not in one of the critical areas described in this section for Type I actions and the construction of utility facilities to serve such establishments

(d) Actions involving individual setback and lot line variances

(e) Agricultural farm management practices including construction, maintenance and repair of farm buildings and structures and land use changes consistent with generally accepted principles of farming

(f) Operation, repair, maintenance or minor alteration of existing structures, land uses and equipment

(g) Restoration or reconstruction of a structure in whole or in part being increased or expanded by less than 50% of its existing size, square footage or usage

(h) Repaving of existing highways not involving the addition of new travel lands

(i) Street openings for the purpose of repair or maintenance of existing utility facilities

(j) Installation of traffic control devices on existing streets, roads and highways other than multiple fixtures on long stretches

(k) Mapping of existing roads, streets, highways, uses, and ownership patterns

(l) Regulatory activities not involving construction or changed land use relating to one individual, business, institution or facilities such as inspections, testing, operating certification or licensing

(m) Sales or surplus government property other than land, radioactive material, pesticides, herbicides, or other hazardous materials

(n) Collective bargaining activities

(o) Operating, expense or executive budget planning, preparation and adoption not involving new programs or major reordering of priorities

(p) Investments by or on behalf of agencies or pension or retirement systems

(q) Actions which are immediately necessary for the protection or preservation of life, health, property or natural resources

(r) Routine administration and management of agency functions not including new programs or major reordering of priorities

(s) Routine license and permit renewals where there is no significant change in preexisting conditions

(t) Routine activities of educational institutions which do not include capital construction

Section 617.16. Confidentiality.

A potential applicant to one or more agencies may, prior to applying for a permit, request a pre-application conference with all agencies that may be involved in determining lead agency. At such conference the potential applicant may identify those elements of the project which are in the nature of trade secrets or information, the nature of which if disclosed to the public or otherwise widely disseminated, would cause substantial injury to the competitive position of the potential applicant's enterprise. A potential applicant may request that such elements be kept confidential in accordance with the provisions of the Freedom of Information Law and other applicable laws.

Section 617.17. Effective date.

This Part shall take effect as prescribed in article 8 of the environmental conservation law, as amended.

Appendix C
The Economics of Environmental Impact Statements Questionnaire

The Economics of EISs

The Economics of Environmental Impact Statements
A Study by The Institute on Man and Science

STUDY CLIENTELE:
State and local governmental agencies presently
formulating, enacting or revising an EIS program

PRIMARY INVESTIGATOR:
Gordon A. Enk
Director, Economic & Environmental Studies
The Institute on Man and Science
Rensselaerville, New York 12147
(518)797-3783

RESEARCH ASSISTANT:
Kathryn Ann Troll
Yale School of Forestry
and Environmental Studies

Please return the questionnaire by:
SEPTEMBER 30, 1976

KEY TO QUESTIONNAIRE

One purpose of the enclosed questionnaire is to clarify the roles of the various agencies responsible for implementing the State Environmental Policy Act (SEPA). Please review the Agency's EIS Role-Descriptions presented below and either check that description which best fits your agency's role or list the agency(ies) in your state fitting the descriptions. After identifying the nature of your agency, please answer all questions in the Green Section as well as those in the Section(s) color-coded to your Agency's Role-Description(s). (Note: More than one description may fit your agency's EIS role. If this is the case, answer all questions appropriate to each suitable agency description.

EIS AGENCY ROLE DESCRIPTIONS

TYPE I: EIS RESPONSIBLE AGENCY ("EIS Agency") is the agency which, because it is charged with developing rules and regulations for statewide implementation of the Act, plays an integral role in receiving all state EISs.
    Is this your agency's responsibility?   Yes____No____
    If yes, answer all questions in the Pink Section
    If no, which state agency(s) is the Responsible Agency_____

TYPE II: REVIEW AGENCY ("Review Agency") is an agency that does not originate statements on projects with significant environmental effect but rather reviews EISs which are submitted by other agencies (federal, state, or local).
    Is this your agency's responsibility?   Yes____No____
    If yes, answer all questions in the Blue Section
    If no, which state agency(s) is the Review Agency_____

TYPE III: PROJECT ORIGINATING AGENCY ("P. O. Agency") is the agency that undertakes projects of the nature that would require an EIS as mandated by the SEPA.
    Is this your agency's responsibility?   Yes____No____
    If yes, answer all questions in the Orange Section
    If no, which state agency(s) is the Project Originating Agency_____

TYPE IV: CLEARINGHOUSE AGENCY ("C/H Agency") is the agency responsible for the review procedures of the SEPA by acting as a collection and distribution center for federal, state, or local EISs.
    Is this your agency's responsibility?   Yes____No____
    If yes, answer all questions in the Beige Section
    If no, which state agency(s) is the Clearinghouse Agency_____

If you have positively identified your agency, please answer all questions in the Green Section on the background of the state EIS program, as well as those questions under any other appropriate Sections. If your agency does not serve any of the EIS functions described above, please clarify the nature of your agency in the space below and return the entire questionnaire.

Thank you.

Could you please send us:

☐ A copy of your SEPA.

☐ A copy of all implementation guidelines and procedural regulations drafted by your agency.

☐ A copy of other agencies' guidelines that are in your possession.

☐ A copy of any pre-EIS environmental assessment forms.

☐ A copy of any report or study done that evaluates the implementation status of the legislation.

☐ A copy of the EIS-Responsible Agency's annual report (e.g., CEQ equivalent report).

☐ A list of county commissions, regional planning agencies, citizen organizations, etc., with which a Project Originating Agency consults in the preparation and review of an EIS.

☐ A flowchart of how the EIS process operates within your agency and between other involved agencies.

☐ A flowchart depicting the organizational hierarchy and EIS responsibility within your agency.

☐ A copy of the FY '75 budget proposal and budget expenditure, and a copy of FY '76 budget proposal with the legislative appropriations, both accompanied by an explanation of budget items requested.

☐ A copy of any handbooks or program agendas for training courses that develop general methods for environmental impact analysis.

Please put a check next to any material that you are able to send to us.
Thank you for your interest and cooperation.

All respondents are to answer Green Section questions (No. 1-14) in order to provide an overview of the organizational impacts of the EIS program in your state.

1. What agencies (state and/or local) are subject to the EIS requirement as mandated by your State Environmental Policy Act (SEPA)?

2. Does your agency have primary responsibility for any other environmental legislation, such as administering a wetlands program?   Yes_____No_____
   If yes, please explain briefly the agency's function in relation to the environmental legislation.

3. Does your agency issue permits on activities that would require an EIS?
                                                        Yes_____No_____
   Is the permit process coordinated with the EIS program (e.g., permit approval must follow 30 days after the review of a draft EIS)? Yes_____No_____
   If no, why not?

4. Since the enactment of the SEPA has there been any organizational restructuring within your agency (e.g., formation of an Office of Environmental Impact Review)?
                                                        Yes_____No_____
   If yes, describe the changes and the major reasons for them. Please note if the changes are directly attributable to the SEPA.

5. Does your agency compile EIS-generated information for any further uses (e.g., developing an understanding of "baseline" ecological conditions)?
                                                        Yes_____No_____
   If yes, please explain the mechanism for compiling the information (e.g., data banks) and its intended use.

6. Within your agency is there a professional interdisciplinary team which coordinates the agency's EIS program? Yes____ No____
   If yes, list the number of individual team members and types of disciplines.

7. In the first year after the enactment of your SEPA (19__), what level of staffing did your agency have for implementing the SEPA?
   inadequate_____ adequate_____ more than adequate_____
   If inadequate, please explain your reasons for judging it to be inadequate, the extent of the staffing, and any efforts made to compensate for this handicap.

8. What level of staffing does your agency presently have for fulfilling the SEPA requirements relevant to your agency?
   inadequate_____ adequate_____ more than adequate
   If inadequate, please explain your reasons for judging it to be so, the extent of the staffing at present and any effort to compensate for this handicap.

9. In order to implement the SEPA, has the agency hired new personnel (not merely reorganized existing personnel)? Yes____ No____
   If yes, please indicate the sequential increase of personnel and the incurred salary cost for the fiscal years following enactment of your SEPA.

	Fiscal Years							
	1st year 19__		2nd year 19__		3rd year 19__		4th year 19__	
Additional Staff	number of new employ.	total annual salary	number of new employ.	total annual salary	number of new employ.	total annual salary	number of new employ.	total annual salary
Administrative								
Research								
Clerical								

10. Does your agency assess the costs related to the SEPA? Yes____ No____
    If yes, please describe briefly the system of cost assessment used.
    If no, why not? Who does do such assessment?

11. Please answer the following two questions by checking (√)the appropriate column
   or listing the date of the drafting (Dr), enactment (En), or/and amendment (Am)
   of the guidelines on the chart.

   Question #1: Which agencies are responsible for the preparation
   of guidelines and procedures for the implementation of your SEPA?
   Question#2 : What is the status (Dr,En, or Am) of your guidelines?

GUIDELINES OR PROCEDURES	AGENCY PROFILE				STATUS		
	EIS Agency	Review Agency	P.O. Agency	C/H Agency	Dr	En	Am
For defining the actions that require pre-EIS environmental assessment							
For defining any pre-EIS environmental assessment process							
For reviewing EIS preparation notices & negative declaration notices							
For determining actions which require an EIS							
For determining which actions can be exempt from the provisions of the Act							
For preparing the EIS (data)							
For defining the scope/content of the draft and final EISs							
For defining the distribution & collection of draft EISs during the review process							
For making evaluations & reviews of a draft EIS							
For holding public hearings and meetings							
For public notification of the draft EIS							
For public notification of the final EIS							
For determining the obligations of two or more branches of government involved in the same project							
For determining the lead agency for an EIS project							
For coordinating the EIS into any permit review process							
For coordinating state and federal cooperation on federal projects of particular state concern							
For determining applicability of the law to projects begun before the enactment of EIS legislation							
For resolving time conflicts w/ other statutes binding on project agencies							
For resolving agency conflicts encountered in the EIS review							
For resolving conflicts of any and all agencies working on the same project							
For integrating projects of a common geographical/environmental problem into broad policy or program EISs							
For reporting potential litigation pursuant to the SEPA							

5

12. Please indicate the number of EISs prepared or reviewed during 1975. The following categories have been provided in the interest of research continuity, but if you wish to add categories or projects, please do so at the bottom of the form. If your agency has access to state totals, please provide the numbers. Thank you.

PROJECT TYPE	Total No. EIS prepared statewide			No. EIS prepared by your agency			No. EIS reviewed by your agency		
	On private projects	On local proj.	On state proj.	On priv. proj.	On local proj.	On state proj.	On priv. proj.	On local proj.	On fed'l proj.
TRANSPORTATION PROJECTS									
Highways and road construction									
Surfacing or paving of roads									
Bridges & crossways									
Other projects(list)									
WATER MGMT PROJECTS									
Wetland filling and draining									
Drainage alteration									
Dams or impoundments									
Other projects(list)									
WASTE MGMT PROJECTS									
Landfills									
Sewage treatment									
Other projects(list)									
MAJOR PHYSICAL REMOVAL AND/OR ALTERATION PROJ.									
Surface excavation of minerals									
Lumbering									
Subsurface mining									
Landscaping									
Other projects(list)									
PUBLIC FACILITY CONSTRUCTION/ADDITIONS									
City & state parks									
Recreation structures									
Public office bldgs.									
Univ. and schools									
Other projects(list)									
COMMERCIAL CONSTR. PROJ.									
Shopping centers									
Other commercial structures									
RESIDENTIAL CONSTR.PROJ.									
Low density res. dev. (4 units/less per acre)									
High density res. dev. (5 units/more per acre)									
INDUSTRIAL CONSTRUCTION									
Industrial parks									
Other projects (list)									
OTHER PROJECTS (identify)									

LITIGATION

13. Have any projects under your SEPA generated litigation in which the EIS was the focus of the controversy?   Yes_____No_____
If yes, please provide the following information.

Names of Cases:_____

_____

_____

Description or name of action challenged:_____

_____

_____

SEPA Issue(s): A) No EIS   _____
                B) Inadequate EIS
                   1) Procedural defect(s)   _____
                   2) Defect(s) in content   _____
                C) Substantive challenge to state
                   action under SEPA   _____
                D) Other procedural SEPA issues
                   (please specify)   _____

14. Did any environmental court litigation influence the drafting and/or enacting of the SEPA?   Yes_____No_____
If yes, please provide the following information.

Names of Cases:_____

_____

_____

Description or name of action challenged:_____

_____

_____

How did this challenge influence the drafting and/or enacting (circle one) of the SEPA?

All of the questions in this Pink Section (No. 15-22) are to be answered only by
representatives of those agencies responsible for developing rules and regulations
for statewide implementation of the SEPA. Responses to the question will be used
in developing 1)a general overview of procedural policies re: SEPA implementation
and 2)a more detailed outline of related administrative costs.

15. What criteria are used by your agency to determine if a project constitutes a "major
action"?

16. What criteria are used by the agency to determine if a project constitutes a
"significant effect"?

17. Since the enactment of the SEPA has there been any organizational restructuring
(e.g., creation of Super-Dept) between the affected state agencies?

Yes_____No_____

If yes, describe the changes, the major reasons for them, and their impacts.

18. What activities has your agency undertaken or planned to undertake to develop general methods (e.g., handbooks, training courses) for environmental impact analysis?

19. What efforts, if any, have been made for public review and/or public involvement in the drafting of the SEPA implementing rules and regulations (e.g., formation of a citizen advisory council)?

20. Does your agency invoke any jurisdictional and/or special responsibilities in reviewing EISs or individual agency guidelines?   Yes____No____
If yes, please explain the nature of the jurisdiction and/or responsibilities.

21. Are there any documents used by your agency to judge the effectiveness and efficiency of the implementation of the SEPA (e.g., EISs, hearing records)?
                                                      Yes____No____
If yes, please describe.

COSTING

22. For the FY 1975 what was the cost that your agency incurred in <u>administering</u> the EIS program and in <u>reviewing</u> state EISs? For each of the following items on the next two pages, please specify (✓) whether the cost is actual (A) or estimated (E):

10.

Budgetary Items: Cost Breakdown	Administrative Costs			Review Costs		
	Costs	A	E	Costs	A	E
Total dollar cost	$					
Is the cost given shown accountable to the SEPA on the agency's budget proposals? Yes____ No____ Percentage represented by the above cost to the agency's total operating budget: _____						
Number of staff persons required to fulfill:						
The administrative responsibilities of the EIS program						
The reviewing responsibilities of incoming EISs						
The clerical workload						
Agency staff hours expended in review of one (typical or average) EIS.						
Please briefly explain the nature of the EIS used in computing the above time estimate: Please mention the range of staff time-high and low-committed to other EISs.						
General (average) salary of staff personnel who expend time reviewing EISs and the rule and regulations pursuant to the SEPA (Give salary range here:_____)	$					
Total cost of hiring consultants for any special review of EISs	$					
Total cost of hiring consultants for special reviews of the implementation status of the EIS legislation	$					
Total cost of hearings on the rules and regulations	$					
Total cost of handbooks or training seminars for implementation of the SEPA	$					
Total cost of equipment, materials and supplies	$					

COST COMPONENTS:

If the costs cannot be broken down in the above terms, please indicate (✓) whether or not the following factors are included in the <u>total dollar cost</u> mentioned at the beginning of this question or whether the factors are not applicable to your agency (N/A):

11.

Budgetary Items: Cost Components		Administrative Costs			Review Costs		
		Yes	No	N/A	Yes	No	N/A
AGENCY STAFF TIME	For the review of incoming EISs (including any and all resource inventories)						
	For the administration of the EIS program						
	For any interagency EIS task force						
	For the clerical workload incurred by the EIS program in your agency						
	For the review of pre-EIS environmental assessments						
	For the review of other state permits and programs						
	For the review of any federal EISs						
	For testimony at any hearing pursuant to the SEPA						
	For other SEPA related work (e.g. publishing annual report)						
	CONTRACTOR STAFF TIME For conducting any special review related to the agency's SEPA implementation efforts						
SUPPLIES	AGENCY STAFF TRAVEL EXPENSES						
	For publication and circulation of all guidelines						
	For supplying miscellaneous information to EIS applicants						
SPECIAL COSTS	For conducting a hearing on any rules and regulations generated pursuant to SEPA						
	For any SEPA training courses or impact assessment handbooks on preparing and reviewing EISs						
	For creation of a data bank or similar information system						
	For litigation pursuant to the SEPA						
	For any other costs incurred by your agency's effort to implement your SEPA (identify)						

All of the questions in this Blue Section (No. 23-28) are to be answered only by representatives of those agencies which are responsible for reviewing EISs submitted by other federal, state, or local agencies. Responses to the questions will be used in developing 1)a general overview of qualitative review criteria and 2)a more detailed outline of the costs involved in the review process.

23. What criteria are used to judge the "adequacy" of an EIS (i.e., meeting all the procedural standards)?

24. What criteria are used to judge the "sufficiency" of an EIS (i.e. addressing all environmental concerns in an objective and informative manner which is useful for the decision-maker)?

25. Does your agency examine the final EIS to see if the Project Originating Agency responded to your comments made on the draft?   Yes____No____
    If yes, please describe the follow-up procedures.

26. Are the review criteria for draft and final statements the same?
                                                        Yes____No____
    If no, what are the differences?

27. Does your agency actively involve any individuals with special expertise (e.g. noise technicians) in reviewing EISs?   Yes____No____
    If yes, please explain the ways in which these experts are used.

COSTING

28. For the FY 1975 what costs has your agency incurred for reviewing EISs pursuant to the SEPA? For each of the following items, please specify (✓) whether the cost is Actual (A) or Estimated (E):

Budgetary Items: Cost Breakdown	Review Costs		
	Costs	A	E
Total dollar cost	$		
Is the cost given above shown accountable to the SEPA on your agency's budget proposals?　　　　　Yes____No____ Percentage represented by cost to the agency's operating budget____			
Number of staff persons required to fulfill the EIS review responsibility			
Agency staff hours expended in review of one EIS (Please explain the nature of the EIS used in computing the above time estimate)			
Average salary of staff personnel who expend time reviewing EISs (salary range:_____)	$		
Cost of hiring consultants for special review	$		
Equipment and material expenses	$		

If the costs cannot be broken down in the above terms, please indicate (✓) whether or not the following factors are included in the total dollar cost mentioned at the beginning of this question, or whether the factors are not applicable to your agency

	Budgetary Items: Cost Components	Yes	No	N/A
STAFF TIME	For review of incoming EISs, including any and all resource inventories			
	For review of any federal EISs			
	For the clerical workload incurred by the EIS program			
	For the administration of the EIS review (drafting guidelines)			
	For any interagency EIS task force			
	For testifying at any hearing pursuant to the SEPA			
	For review of pre-EIS environmental assessments			
	For review of other state permits and programs			
	For review of other state permits and programs			
	CONTRACTOR STAFF TIME			
SUPPLIES	AGENCY STAFF TRAVEL EXPENSES			
	For publication and circulation of all guidelines			
	For supplying miscellaneous information to EIS applicants			
SPECIAL COSTS	For creation of a data bank or similar information system			
	For any SEPA training courses or impact assessments handbooks for reviewing EISs			
	For any other costs incurred by your agency's effort to implement your SEPA (identify)			

All questions in this Orange Section (No. 29-37) are to be answered only by representatives of those agencies undertaking projects that would require an EIS as mandated by the SEPA. Responses to the questions will be used in developing 1) an overview of the role of citizen participation in the EIS process, and 2) a profile of the EIS decision-makers, 3) an outline of the components of the decision-making process, and 4) a more detailed look at costs related to the preparation, review and administration of EISs.

29. To what extent has your agency sought citizen participation on a regular basis in the planning, preparing and reviewing of the agency's draft and final EISs? Please indicate (✔) in the appropriate column the means of public solicitation and/or public involvement for each stage of the EIS process.

		EIS STAGE						
		1	2	3	4	5	6	7
PUBLIC SOLICITATION	Open public file							
	Bulletin board notice							
	Publication in some form of EIS Monitor							
	Newspaper notice							
	Personally informing public organizations							
	Informing public organizations w/ special interests							
PUBLIC INVOLVEMENT	Public meeting							
	Formal public hearing							
	Citizen accompaniment on some site visits							
	Citizen consultation by some interagency task force							
	Reviewing written comments from individuals							
	Reviewing written comments from public organizations							

KEY TO EIS STAGES:

1. Decision to include a project in the EIS process
2. Gathering of data
3. Development of alternatives
4. Review of draft EIS
5. Decision on proposed action
6. Review of final EIS
7. Decision on project disposition or decision to grant a permit on an activity in the EIS

30. Has your agency participated in any "joint statements" with another involved agency on any EIS project?   Yes_____No_____
    If yes, please explain the participatory roles of each agency.

31. What criteria are used by your agency to determine if a proposed project constitutes a "major action"?

32. What criteria are used by the agency to determine if a proposed project constitutes a "significant effect"?

33. Who makes the go/no-go decision on a project and by what means? Please state the means of making the decision (e.g. granting a specific activity permit or blanket approval of the project as the EIS) in the blank following the description of all those decision-makers who have a final say on a project.

<u>Decision-Makers</u>   <u>Means of Imposing the Decision</u>

Agency Preparers of the EIS   _____

Agency Reviewers of the EIS   _____

Permit Grantors   _____

Commissioner or Director of the
  entire agency   _____

Director of the in-house EIS office   _____

Interagency Council or Committee   _____

34. What project documents are used to present environmental analyses to the
    <u>decision makers</u> (e.g., hearing record, EIS, decision memoranda)?   16.

35. Does your agency monitor projects on which an EIS has been written and approved?
                                                              Yes_____ No_____
    If yes, please explain the efforts made to:
    A) ensure that all proposed mitigating measures have been met:

    B) determine the adequacy of the data and impact analysis given in the EIS:

36. What activities has your agency undertaken (or planned to undertake) to develop
    general methods for environmental impact analysis (e.g., handbooks, training
    courses, etc.)?

COSTS

37. For the FY 1975 what is the cost that your agency has incurred in preparing, reviewing, and administering the EIS as mandated by the SEPA? Please indicate (√) whether the cost of each item below is actual (A) or estimated (E):

Budgetary Items: Cost Breakdown	Cost	A	E
Total dollar cost of the EIS program as implemented within your agency	$		
Is the cost given above accountable to the SEPA on the agency's budget proposal Yes____No____ Percentage represented by the above cost to the agency's operating budget for 1975_____			
Cost for staff persons required to fulfill:			
the administrative responsibilities of the EIS program	$		
the research effort in preparing EISs	$		
the review responsibility of incoming EISs	$		
the clerical workload	$		
Agency staff hours expended to prepare, review and administer one (typical or average) of the agency's EIS			
Please explain briefly the nature of the EIS used in computing the above time estimate). . Please comment on range (high & low) of staff time committed to other EISs			
Average salary of staff personnel who expend time preparing and reviewing EISs (salary range:_____)	$		
Cost of hiring consultants for preparing an EIS	$		
Cost of hearings on EIS projects	$		
Cost of printing and distributing the draft EIS and the final EIS	$		

If the costs cannot be broken down in the above terms, please indicate (✓) whether or not the following factors are included in the total dollar cost mentioned at the beginning of this question, or whether the factors are not applicable (N/A) to your agency.

18.

Budgetary Items: Cost Components		Yes	No	N/A
STAFF TIME	For the review of incoming EISs			
	For the preparation of EISs			
	For the administration of the EIS program (drafting guidelines)			
	For any interagency EIS task force			
	For the clerical workload incurred by an EIS program			
	For resource inventories necessary for the preparation and review of EISs			
	For review of pre-EIS environmental assignments			
	For testifying at hearings pursuant to the SEPA			
	For review of other state permits and programs			
	For review of any federal EISs			
CONTRAC-TOR'S TIME	For the preparation of EISs			
	For conducting any special review of the agency's program			
	STAFF TRAVEL EXPENSES			
SUPPLIES/MATERIALS	For producing the EIS			
	For printing and circulating drafts and final EISs			
	For publication and circulation of all guidelines			
	For supplying miscellaneous information to EIS applicants			
	For any other costs relevant to the review of your SEPA (identify)			
SPECIAL COSTS	For conducting a hearing on agency projects			
	For any extraordinary EIS project cost			
	For SEPA training courses or impact assessments handbooks on preparing and reviewing EISs			
	For creation of a data bank or similar information system			
	For any other costs incurred by your agency's efforts at SEPA implementations (identify)			

All questions in the Beige Section (No. 38-45) are to be answered only by representatives of Clearinghouse Agencies, those agencies which have responsibility for the review procedures of the SEPA and act as a collection and distribution center for federal, state, and/or local EISs. Responses to the questions will be used in developing 1) an overview of the role of the Clearinghouse Agency, and 2) a more detailed outline of the costs incurred by such agencies during the collection, review, and distribution of EISs.

38. To which agencies, organizations, individuals are the draft and final state EISs distributed?

39. Does your agency collect the various agency comments on a proposed project?
Yes_____No___
If yes, by what means? If no, why not? Who does collect the comments?

40. Does your agency prepare a consolidated uniform state response to a proposed project or return comments verbatim to the Project Originating Agency? Please give the rationale behind your agency's handling of this matter.

41. Does your agency participate in reviewing the EIS

       for scope and content   Yes_____ No_____
       for compliance with all procedural requirements   Yes_____ No_____
       to express an opinion on project disposition   Yes_____ No_____

Why or why not?

42. Does your agency solicit public response or notify the public before collecting
and returning the comments to the Project Originating Agency?
Yes_____ No_____
If yes, by what means?

43. Before distributing the DEISs does your agency screen for referral according
to type of action (minor, major, etc.),   the need for unique review (technical
specialist, affected commission, etc.),   or the need for review by "concerned"
agencies?

44. Does your agency send the collection of comments or consolidated reports to any
other branch of government in the state?
Yes_____ No_____
If so, where and to whom: to which agencies and/or individuals?

45. For the FY 1975, what is the cost that your agency incurred in collecting, distributing and reviewing EISs? For each item listed below, please indicate with a check (✓) whether the cost figure is actual (A) or estimated (E).

Budgetary Items: Cost Breakdown	Cost	A	E
Total dollar cost	$		
Is the cost given above accountable to the agency's operating budget? Yes_____ No_____ Percentage of the above figure to the agency's operating budget:_____			
The number of staff persons required to fulfill the administration of clearinghouse function for EISs			
the clerical workload			
Average salary of staff members who spend time collecting, circulating and distributing EISs (Note: Enter salary range here:_____)	$		
Equipment and material expenses	$		

If the costs cannot be broken down in the above terms, please indicate (✓) whether or not the following factors are included in the total dollar cost mentioned at the beginning of this question or whether the factors are not applicable (N/A) to your agency.

	Budgetary Items: Cost Components	Yes	No	N/A
STAFF TIME	For the collection and circulation of the EISs			
	For the review of the EISs			
	For the clerical workload necessitated by the EIS program			
	For the circulation and review of state permits and programs			
	For the circulation and review of federal EISs			
	For the administration of the EIS program in your agency (e.g., drafting in-house EIS procedures, etc.)			
	Contractor staff time (identify purpose of time on back)			
SUPPLIES	For the circulation of EIS copies and comments			
	For the publication and circulation of all guidelines and procedures			
	For supplying miscellaneous information to EIS applicants			
SPECIAL COSTS	For creation of a data bank or similar information system			
	For any other costs incurred by your agency's effort to implement your SEPA (identify:_____)			

Appendix D
Agencies and Personnel Contacted

CALIFORNIA

Questionnaire sent to:

Norman E. Hill
Assistant to the Secretary
The Resources Agency of California
Resources Building
1416 Ninth St.
Sacramento, CA 95814

Personal contact with:

Norman Hill
(see above)

Donald Meixner,
Jake Angel, and
John McClurg
Dept. of Water Resources
Resources Bldg.
1416 Ninth St.
Sacramento, CA 95814

Thomas Willoughby
Chief Consultant
Assembly Committee on
  Resources, Land-Use
  and Energy
State Capitol
Sacramento, CA 95814

Scott Warner and
Robert Remen
Office of Planning and
  Research
1400 Tenth Street
Sacramento, CA 95814

William Blackmer
Assistant Chief
Office of Environmental
  Planning
Department of Transporta-
  tion (CalTrans)
1120 N St.
Sacramento, CA 95814

Alcides Freitas
Environmental Coordinator
Environmental Impact Sect.
County of Sacramento
827 7th St., Rm. 301C
Sacramento, CA 95814

CONNECTICUT

Questionnaire sent to:

Jonathan T. Clapp
Environmental Analyst
Office of Planning and Coordination
Connecticut Dept. of Environmental Protection
State Office Building
Hartford, CT 06115

Personal contact with:

    Jonathan Clapp
    (above)

    Gregory Sharp
    Director, Information & Education
    Dept. of Environmental Protection
    State Office Bldg., Rm. 114
    Hartford, CT 06115

HAWAII

Questionnaire sent to:

Richard Marland, former Director, Office of Environmental Quality Control (as of 1/78) 550 Halekawila St. Room 301 Honolulu, HI 96813	T. Harano Chief Land Transportation Facilities Div. Hawaii Dept. of Transportation 869 Punchbowl St. Honolulu, HI 96813

Personal contact with:

    Richard O'Connell
    Director
    Office of Environmental Quality Control
    550 Halekawila St., Rm. 301
    Honolulu, HI 96813

INDIANA

Questionnaire sent to:

    Ralph Pickard
    Technical Secretary
    Indiana Environmental Management Board
    State Board of Health
    1330 W. Michigan St.
    Indianapolis, IN 46206

Personal contact with: Same

MARYLAND

Questionnaire sent to:

Joseph Knapp
Office of the Secretary
Maryland Department of
  Natural Resources
Towes State Office Bldg.
580 Taylor Ave., C-4
Annapolis, MD 21401

Richard Krolak
Environmental Assessment
Bureau of Project Planning
Maryland State Highway Ad-
  ministration
Department of Transporta-
  tion
300 W. Preston St.
Baltimore, MD 21201

Personal contact with:

Joseph Knapp
(above)

Warren Hodges or
James McConnaughay
State Clearinghouse
Maryland Department of
  State Planning
301 W. Preston St.
Baltimore, MD 21201

MASSACHUSETTS

Questionnaire sent to:

William Hicks, Chief Planner
Massachusetts Executive
Office of Environmental Affairs
100 Cambridge St.
Boston, MA 02202

Personal contact with:

William Hicks (above)

John Hurley
Environmental Engineer
Dept. of Public Works
100 Nashua St.
Boston, MA 02114

Marge Luening
Massachusetts Housing
  Finance Agency
Old City Hall
45 School St.
Boston, MA 02108

Richard Bates
Department of Environ-
  mental Quality En-
  gineering
100 Cambridge St.
Boston, MA 02202

David Russell
Engineer
Department of Public
  Utilities
Boston, MA 02114

MICHIGAN

Questionnaire sent to:

Terry L. Yonker, former
Executive Secretary
Michigan Environmental
  Review Board
Dept. of Management
  and Budget
P.O. Box 30026
Lansing, MI 48909

Mr. Adams, Chief
Environmental & Community
  Factors Section
Michigan Dept. of State
  Highways & Transporta-
  tion
P.O. Box 30050
425 W. Ottawa St.
Lansing, MI 48909

Personal contact with:

Terry Yonker
(above)

Boyd Kinzley
Executive Secretary
Michigan Environmental
  Review Board
P.O. Box 30026
Lansing, MI 48909

MINNESOTA

Questionnaire sent to:

John L. Robertson
Assistant Director
Program Activities
Minnesota State Plng.
  Agency
Environmental Quality
  Council
Capitol Square Bldg.
550 Cedar St.
St. Paul, MN 55101

Dale McMichael
Coordinator
Office of Environmental
  Analysis
Minnesota Pollution
  Control Agency
1935 W. Country Rd., B-2
Roseville, MN 55113

Personal contact with:

Dale McMichael (above)

John Robertson (above)

Nancy Onkke
State Planning Agency
550 Cedar St.
St. Paul, MN 55101

Charles Kenow
Coordinator, EIS Prog.
State Planning Agency
Environ. Quality Council
550 Cedar St.
St. Paul, MN 55101

Robert Herbst
Dept. of Natural Resources
Centennial Office Bldg.
658 Cedar St., 3rd Fl.
St. Paul, MN 55155

MONTANA

Questionnaire sent to:

> Wayne Wetzel
> Environmental Coordinator
> Montana Department of Natural Resources
>   and Conservation
> 32 South Erving
> Helena, MT 59601

John Reuss
Montana Environmental
  Quality Council
Box 215
Capitol Station
Helena, MT 59601

Duane Noel
Ecological Researcher
Montana Environmental
  Quality Council
Capitol Station
Helena, MT 59601

Personal contact with: same

NEW JERSEY

Questionnaire sent to:

Larry Schmidt
Office of Environmental
  Review
New Jersey Department of
Environmental Protection
P.O. Box 1390
Trenton, NJ 08625

Frank Winters
Project Engineer
Bureau of Environmental
  Analysis
New Jersey Department of
  Transportation
1035 Parkway Ave., Rm.20100
Trenton, NJ 08625

Personal contact with: same

NEW YORK

Questionnaire sent to:

> Allen F. Davis
> Chief, Project Review Section
> Office of Environmental Analysis
> NYS Department of Environmental Conservation
> 50 Wolf Road
> Albany, NY 12233

Personal contact with:

Allen Davis
(above)

Peter Buttner
Director
Environmental Management
NYS Parks & Recreation
Empire State Plaza
Albany, NY 12223

Bernie Cobb
Division of Budget
NYS Department of Trans-
    portation
State Office Campus
Building #5
Albany, NY 12226

Terence Curran, Director
Office of Environmental
    Analysis
NYS Department of Environ-
    mental Conservation
50 Wolf Road
Albany, NY 12233

Pat Grady and
Michael Morandi
Office of Environmental
    Analysis
NYS Department of Environ-
    mental Conservation
50 Wolf Road
Albany, NY 12233

Keith Smith
Environmental Impact Sect.
NYS Department of Trans-
    portation
State Office Campus
Bldg. #5-Rm. 524
Albany, NY 12226

NORTH CAROLINA

Questionnaire sent to:

> D. Keith Whitenight or
> Thayer Broili
> Department of Natural & Economic Resources
> P.O. Box 27687
> Raleigh, NC 27611

Personal contact with:

D. Keith Whitenight and
Thayer Broili
(above)

Byron O'Quinn
Environmental Planning
    Engineer
Department of Transpor-
    tation
Raleigh, NC 27611

Steven Gluckman
Archaeology Section
NC Department of Cultural
    Resources
Raleigh, NC 27611

SOUTH DAKOTA

Questionnaire sent to:

Harold Lenhart
Policy Analyst
South Dakota Department
  of Environmental
  Protection
Jos Foss Building
Pierre, SD 57501

Ed McGuire
Executive Policy Aide III
SD Office of Executive
  Management
State Planning Bureau
State Capitol
Pierre, SD 57501

Personal contact with: same

TEXAS

Questionnaire sent to:

Albert D. Schutz
Planner
Natural Resources Sect.
Office of the Governor
Budget & Planning Office
Executive Office Bldg.
411 W. 13th St.
Austin, TX 78701

R. L. Lewis
Chief Engineering of
  Highway Design
Texas Dept. of Highways &
Public Transportation
Austin, TX 78701

Henry E. Sievers, P.E.
Abatement Requirements
  & Analysis
Standards & Regulation
  Program Division
Texas Air Control Board
8520 Shoal Creek Blvd.
Austin, TX 78758

Personal contact with: same

UTAH

Questionnaire sent to:

Chauncey G. Powis
Environmental Coordinator
Office of the Utah State Planning Coordinator
118 State Capitol
Salt Lake City, UT 84114

Personal contact with: same

VIRGINIA

Questionnaire sent to:

Reginald F. Wallace
Environmental Impact Statement Coordinator
Virginia Council on the Environment
903 Ninth Street Office Building
Richmond, VA 23219

J. Steven Griles
Virginia Department of
Conservation & Economic
  Development
1100 State Office Bldg.
Richmond, VA 23219

Dennis Gilbert
Virginia Department of
Highways & Transportation
1401 E. Broad St.
Richmond, VA 23219

Personal contact with: same

WASHINGTON

Questionnaire sent to:

Dennis Lundblad and
T. L. Elwell
Environmental Review
Department of Ecology
Olympia, WA 98504

Personal contact with:

Dennis Lundblad and
T. L. Elwell
(above)

Bernie Chaplin
Project Planning Super.
Department of Transpor-
  tation
Highway Administration
  Bldg.
Olympia, WA 98504

Chuck Mize
Washington Association
  of Cities
Seattle, WA

Stan Springer
Supervisor
Environmental Review
Dept. of Ecology
Olympia, WA 98504

James Williams
Planner
Washington State Associa-
  tion of Counties
6730 Martin Way, N.E.
Olympia, WA 98506

WISCONSIN

Questionnaire sent to:

James Sinopoli
Environmental Planner
Wisconsin Dept. of Ad-
 ministration
State Planning Office
1 W. Wilson St.
Madison, WI 53702

James R. Huntoon
former Director
Bureau of Environmental
 Impact
Wisconsin Dept. of Natural
 Resources
Box 450
Madison, WI 53701

Personal contact with:

James Sinopoli
(above)

Caryl Terrell
State WEPA Coordinator
Office of State Plng.
 and Energy
State Capitol
Madison, WI 53702

Howard Druckenmiller
Director
Bureau of Environmental
 Impact
Wisconsin Department of
 Natural Resources
Box 450
Madison, WI 53701

For Product Safety Concerns and Information please contact our EU
representative GPSR@taylorandfrancis.com
Taylor & Francis Verlag GmbH, Kaufingerstraße 24, 80331 München, Germany

www.ingramcontent.com/pod-product-compliance
Lightning Source LLC
Chambersburg PA
CBHW060756220326
41598CB00022B/2449

9 780367 171834